D1797478

Computer Architecture and Design Methodologies

Series editors

Anupam Chattopadhyay, Noida, India
Soumitra Kumar Nandy, Bangalore, India
Jürgen Teich, Erlangen, Germany
Debdeep Mukhopadhyay, Kharagpur, India

Twilight zone of Moore's law is affecting computer architecture design like never before. The strongest impact on computer architecture is perhaps the move from unicore to multicore architectures, represented by commodity architectures like general purpose graphics processing units (gpgpus). Besides that, deep impact of application-specific constraints from emerging embedded applications is presenting designers with new, energy-efficient architectures like heterogeneous multi-core, accelerator-rich System-on-Chip (SoC). These effects together with the security, reliability, thermal and manufacturability challenges of nanoscale technologies are forcing computing platforms to move towards innovative solutions. Finally, the emergence of technologies beyond conventional charge-based computing has led to a series of radical new architectures and design methodologies.

The aim of this book series is to capture these diverse, emerging architectural innovations as well as the corresponding design methodologies. The scope will cover the following.

Heterogeneous multi-core SoC and their design methodology
Domain-specific Architectures and their design methodology
Novel Technology constraints, such as security, fault-tolerance and their impact on architecture design
Novel technologies, such as resistive memory, and their impact on architecture design
Extremely parallel architectures

More information about this series at http://www.springer.com/series/15213

Jawad Haj-Yahya · Avi Mendelson
Yosi Ben Asher · Anupam Chattopadhyay

Energy Efficient High Performance Processors

Recent Approaches for Designing Green High Performance Computing

 Springer

Jawad Haj-Yahya
School of Computer Science and
  Engineering
Nanyang Technological University
Singapore
Singapore

Avi Mendelson
Department of Computer Science
Technion—Israel Institute of Technology
Haifa
Israel

Yosi Ben Asher
Department of Computer Science
University of Haifa
Haifa
Israel

Anupam Chattopadhyay
School of Computer Science and
  Engineering
Nanyang Technological University
Singapore
Singapore

ISSN 2367-3478   ISSN 2367-3486   (electronic)
Computer Architecture and Design Methodologies
ISBN 978-981-10-8553-6   ISBN 978-981-10-8554-3   (eBook)
https://doi.org/10.1007/978-981-10-8554-3

Library of Congress Control Number: 2018933463

© Springer Nature Singapore Pte Ltd. 2018
This work is subject to copyright. All rights are reserved by the Publisher, whether the whole or part
of the material is concerned, specifically the rights of translation, reprinting, reuse of illustrations,
recitation, broadcasting, reproduction on microfilms or in any other physical way, and transmission
or information storage and retrieval, electronic adaptation, computer software, or by similar or dissimilar
methodology now known or hereafter developed.
The use of general descriptive names, registered names, trademarks, service marks, etc. in this
publication does not imply, even in the absence of a specific statement, that such names are exempt from
the relevant protective laws and regulations and therefore free for general use.
The publisher, the authors and the editors are safe to assume that the advice and information in this
book are believed to be true and accurate at the date of publication. Neither the publisher nor the
authors or the editors give a warranty, express or implied, with respect to the material contained herein or
for any errors or omissions that may have been made. The publisher remains neutral with regard to
jurisdictional claims in published maps and institutional affiliations.

Printed on acid-free paper

This Springer imprint is published by Springer Nature
The registered company is Springer Nature Singapore Pte Ltd.
The registered company address is: 152 Beach Road, #21-01/04 Gateway East, Singapore 189721, Singapore

Preface

There is a growing demand worldwide for resource conservation in all aspects of daily life. Utilizing resources effectively, improving environmental performance, and defending global warming are in priority on the list of global challenges that must be addressed urgently. Governments and business associations have introduced a range of initiatives to address environmental challenges, particularly global warming and energy use. Business associations have mainly developed initiatives to reduce energy costs and to demonstrate corporate social responsibility. And this is correct for computing systems.

Green computing is a large and increasing area. The need for saving energy has become a top priority in almost all segments of the IT market. In addition, the need for power efficiency has become a critical factor in the design of High-Performance Computing (HPC). The high technology industries are heavily investing in research, development, and manufacturing of energy-efficient computing systems in all ranges of applications, from embedded systems to enterprise data center.

Designing energy-efficient devices has a great return on investment, whether it is to create power-aware embedded system, extending the battery life of portable devices (such as mobile phones, tablets, or PCs) or saving cost in cooling data centers.

This book is intended for system, hardware, and software engineers, helping them understand energy-aware computing, and provide them with hardware and software methodologies and techniques for power and energy savings that can be used when designing computing solutions, so they can contribute to the global effort of building green computing systems.

Recently, many initiatives and efforts investing in energy-efficient computing in many areas, here we share part of them.

The Green Data Center

Data centers are found in all economic sectors, as they provide computational infrastructure for a wide range of applications. It is expected that data centers should always be available with their data. The need for power efficiency has become a critical factor in the design of data centers. In cloud computing, data centers gain popularity as an effective platform for scheduling of resources and hosting cloud applications. A tremendous amount of energy is consumed by these data centers, which leads to high operational costs and contributes toward carbon footprints to the environment.

Nearly half of the data center energy is spent on cooling. Recently, a project held by the National Snow and Ice Data Center (NSIDC) called Green Data Center project. The project funded by the National Science Foundation (NSF) in the United States with additional support from NASA. The project shows that there is significant opportunity to improve the efficiency of data centers. The heart of the design includes new cooling technology that uses 90% less energy than traditional air conditioning, and an extensive rooftop solar array that results in a total energy savings of 70%. The Green Data Center design takes advantage of Colorado's arid climate, using a new technology called indirect evaporative cooling. These units, cool by blowing air over water, using less energy than compressors. Unlike traditional evaporative cooling, indirect evaporative cooling does not add humidity to the room, maintaining the dry environment that computers need. During much of the year, the system cools the data center by pulling in and filtering outdoor air. On hot days, the new cooling units automatically step in. The solar array feeds energy back into the electrical grid, further reducing the center's net carbon footprint.

Heterogeneous Computing

Today's computer systems have high performance demand while they need to maintain low energy consumption. Creating a system that can supply high performance while keeping energy low is a huge challenge to System on Chip (SoC) architects. Recently, heterogeneous computing has been implemented in high-performance computing system in order to raise the energy efficiency.

One of these architectures is the ARM big.LITTLE heterogeneous computing architecture developed by ARM Holdings, coupling relatively battery-saving and slower processor cores (LITTLE) with relatively high-performance and power-hungry ones (big). The architecture is pairing a high-performance Cortex-A15 MPCore and a low power Cortex-A7 processors. Big.LITTLE processing allows devices to select the right processor for the right task, based on performance requirements. Importantly, this dynamic selection is transparent to the application software or middleware running on the processors.

This technology was implemented in many SoCs that are used for mobile devices, such as Samsung Exynos 9 Octa (8895 model) which is used inside Samsung Galaxy devices.

Connected Standby

Connected Standby is a feature used in laptops, tablets, and smartphones in order to reduce energy consumption when the device is fully idle, while maintaining user experience on instant-on and responsiveness. At first, only ARM-based devices (iPad, Android tablets, and smartphones) had this feature, and later on, windows had collaborated with Intel in order to support this feature at Intel-based laptops (Core- and Atom-based); today, the feature is becoming more common in portable devices.

A PC with Connected Standby enters to this state when the device is idle instead of entering into Sleep and Hibernate states. This allows to keep the processor ready for resuming the execution (instant-on) while keeping low power consumption.

Book Structure

This book is divided into five chapters; each chapter covers different aspects of the modern processor's power management.

Chapter 1: Introduction
Chapter 1 introduces the concept of green computing in face of the power wall in portable consumer electronics. It envisions the structure and necessary components for power management of such a system and identifies power management blocks as the bottleneck for overall efficiency improvement. As an introduction, it also describes the uniqueness of power management design and the rationale behind the study thereof.

Chapter 2: Dynamic Optimizations for Energy Efficiency
For the deep understanding of how the energy is distributed and how energy optimizations are applied at modern processor, Chap. 2 covers the main methods that optimize certain processor features for energy efficiency.

Chapter 3: Power Modeling at High-Performance Computing Processors
Central to saving power and energy at modern processor, it is essential to be able to model the power at these systems. In Chap. 3, we explain the importance of power modeling and present an accurate tool that breaks down the power consumption of modern processors at sub-processor core/thread granularity on real systems.

Chapter 4: Compiler-Directed Energy Efficiency
Chapter 4 focuses on how software combined with system features can affect energy efficiency at modern processors. It describes how compilation techniques

along with hardware and firmware changes can reduce the energy consumption of the platform.

Chapter 5: Static Power Modeling for Modern Processor
Power and energy estimation tools are essential tools that are used by system designers, software developers, and compiler developers to raise the energy efficiency of their designs. Chapter 5 presents the static estimation and analyses of energy and power of programs. We demonstrate how these tools can be used to optimize the power and energy of programs at compile time by choosing compiler flags that minimize the energy or power of the program.

Singapore, Singapore   Jawad Haj-Yahya
Haifa, Israel   Avi Mendelson
Haifa, Israel   Yosi Ben Asher
Singapore, Singapore   Anupam Chattopadhyay

Acknowledgements

We would like to thank technical and content reviewers who significantly improved the quality of the book. This includes **Ahmad Yasin**, **Efraim Rotem**, and **Ran Ginosar**.

We thank Intel for providing access to the computer systems and simulators data as well as for **Tal Katz** for lab support.

Thanks to our editors and the people at Springer for their support and patience. Finally, the authors would like to thank their families for their support while the authors spent nights, weekends, and vacations to make this book happen.

Contents

Chapter 1
Power Management of Modern Processors

1.1 Introduction

The trend toward global warming is virtually certain: According to an ongoing temperature analysis conducted by scientists at NASA's Goddard Institute for Space Studies (GISS) [1], the average global temperature on Earth has increased by about 0.8 °C (1.4 °F) since 1880. Two-thirds of the warming has occurred since 1975, at a rate of roughly 0.15–0.20 °C per decade. There is a strong consensus among scientists that this is a result of the emission of greenhouse gases, preeminent of which is CO_2, mainly originating from fossil fuel burning. Aside from the negative environmental impact, fossil fuels are nonrenewable resources that take millions of years to form, but are now depleting at a much faster speed than new ones are being made. As a result, the world is facing an unprecedented challenge to switch from fossil fuel, carbon emission driven growth to an eco-friendly, low-carbon economy to meet the energy need as well as mitigating the environmental consequences [2].

Next to the replacement of fossil fuels by alternatives, such as solar, wind, geothermal, biomass, and so forth, there is also increasing awareness that scrutiny of existing technology and applications can have an immediate and profound impact [3]. Energy efficiency, which is to reduce the amount of energy required to provide products and services, is often considered as the parallel pillar to renewable energy (RE) in a sustainable energy economy [4]. Parallel to the large-scale energy grand challenge, portable and battery-powered products are facing a similar power crisis: consumers, who appreciate the multimedia experience and ubiquitous connectivity in an increasingly compact size, are unwilling to sacrifice and even demand longer battery life for new generations of devices.

This book, therefore, is focused on improving the energy efficiency of electronic products, especially portable and battery-powered ones, using advanced power management and Very-Large-Scale Integration (VLSI) design techniques, which will not only effectively reduce the power consumption and extend the battery life for current consumer products, but also enable and assist future clean energy

© Springer Nature Singapore Pte Ltd. 2018
J. Haj-Yahya et al., *Energy Efficient High Performance Processors*,
Computer Architecture and Design Methodologies,
https://doi.org/10.1007/978-981-10-8554-3_1

generation and distribution, leading to a new era of "Green Electronics" in the future renewable energy economy.

Power consumption in electronic products has become a critical issue that affects the further development of processors. High-power consumption adds challenges to the system design due to the effects of high thermal output, high current requirements, and battery life and electricity costs. In order to raise the awareness, Green500 [5, 6] lists the most energy-efficient supercomputers twice a year according to the ratio of performance and power consumption (as opposed to pure performance ranking). Furthermore, quantifying the power consumption of individual applications is a critical component for software-based power capping, scheduling, and provisioning techniques in modern datacenters [7]. Fine-grain power breakdown is essential for other fronts like power and performance optimization guidelines.

To show the scope of the energy consumption in computer systems, we will take the data centers as an example. Energy efficiency of data centers has attained a key importance in recent years due to its high economic, environmental, and performance impact. First, data centers have high economic impact due to multiple reasons. A typical data center may consume as much energy as 25,000 households. Data center spaces may consume up to 100–200 times as much electricity as standard office space [8]. Furthermore, the energy costs of powering a typical data center double every 5 years [9]. Therefore, with such increase in electricity use and rising electricity costs, power bills have become a significant expense for today's data centers [8, 10]. In some cases, power costs may exceed the cost of purchasing hardware [11]. Second, data center energy usage creates a number of environmental problems [12, 13]. For example, in 2005, the total data center power consumption was 1% of the total United States power consumption [14]. In 2010, the global electricity usage by data centers was estimated to be between 1.1 and 1.5% of the total worldwide electricity usage [15], while in the US the data centers consumed 1.7–2.2% of all US electrical usage [16]. Van Heddeghem et al. [17] have found that data centers worldwide consumed about 270 TWh of energy in 2012 and this consumption had a Compound Annual Growth Rate (CAGR) of 4.4% from 2007 to 2012. Due to these reasons, data center energy efficiency is now considered as a chief concern for data center operators, ahead of the traditional considerations of availability and security. Finally, even when running in the idle mode, servers consume a significant amount of energy. Large savings can be made by turning off these servers. This and other measures such as workload consolidation need to be taken to reduce data center electricity usage. At the same time, these power saving techniques reduce system performance, pointing to a complex balance between energy savings and high performance.

The energy consumed by a data center can be broadly categorized into two parts [18]: energy use by IT equipment (e.g., servers, networks, storage, etc.) and usage by infrastructure facilities (e.g., cooling and power conditioning systems). The amount of energy consumed by these two subcomponents depends on the design of the data center as well as the efficiency of the equipment. For example, according to the statistics published by the Infotech group, the largest energy consumer in a

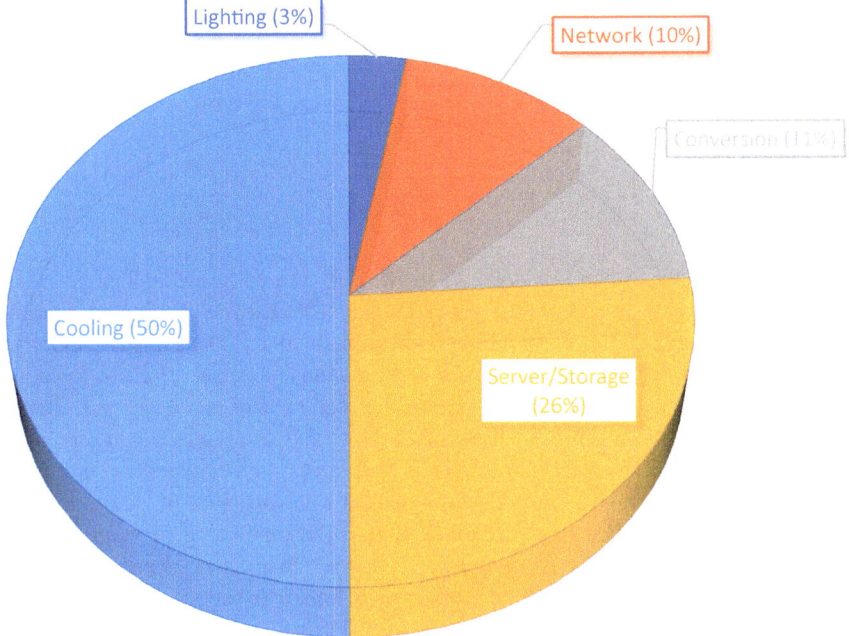

Fig. 1.1 A breakdown of energy consumption by different components of a data center [19]

typical data center is the cooling infrastructure (50%) [17, 19], as shown in Fig. 1.1, while servers and storage devices (26%) rank second in the energy consumption hierarchy. Note that these values might differ from data center to data center [20].

1.1.1   Power Versus Energy

To understand power and energy management mechanisms, it is essential to clearly distinguish the basic terms. We explain the terms of joule, watt, and watt-hour.

Electric current is the flow of electric charge measured in amperes (Amps or simply denoted with A). Amperes define the amount of electric charge transferred by a circuit per second. Power and energy can be defined in terms of work that a system performs. Power is the rate at which the system performs the work, while energy is the total amount of work performed over a period of time. In other words, power can be shown as the *rate* in which energy is consumed.

Power and energy are measured in *watts* (W) and joules (J) respectively, other metrics are used for energy, such as watt-hour (Wh). Work is done at the rate of one watt when one ampere is transferred through a potential difference of one volt.

A kilowatt-hour (kWh) is the amount of energy equivalent to a power of 1 kW (1000 W) running for 1 h. Formally, power and energy can be defined as in (1.1) and (1.2).

$$P = \frac{W}{T} \tag{1.1}$$

$$E = P \cdot T, \tag{1.2}$$

where P is power, T is a period of time, W is the total work performed in that period of time, and E is energy. It is very important to differentiate between the energy and the power, because reduction of the power consumption does not always reduce the consumed energy. For example, the power consumption can be decreased by lowering the CPU performance, however, in this case, a program may require longer time to complete its execution consuming the same or more amount of energy. On one hand, reduction of the peak power consumption will result in decreased costs of the infrastructure provisioning, such as costs associated with capacities of Uninterruptible Power Supply (UPS), Power Distribution Unit (PDU), power generators, cooling system, and power distribution equipment. On the other hand, decreased energy consumption will lead to reduction of the electricity bills. The energy consumption can be reduced temporarily using Dynamic Power Management (DPM) techniques or permanently applying Static Power Management (SPM). DPM utilizes knowledge of the real-time resource usage and application workloads to optimize the energy consumption. However, it does not necessarily decrease the peak power consumption. In contrast, SPM includes the usage of highly efficient hardware equipment, such as CPUs, disk storage, network devices, UPS, and power supplies. These structural changes usually reduce both the energy and peak power consumption.

1.1.2 Energy Efficiency

Energy efficiency means using less energy to provide the same service. For example, a compact fluorescent light bulb is more efficient than a traditional incandescent light bulb as it uses less electrical energy to produce the same amount of light. Similarly, an efficient boiler takes less fuel to heat a home to a given temperature than a less efficient model.

Electrical devices are rated by the amount of energy they consume while in operation and while idle. For example, a 40 W incandescent (traditional) light bulb consumes 40 J of energy per second and produces about 550 Lumens (denoted by *lm*—a measure of the total amount of visible light to the human eye from a lamp or light source). While on the other hand, Light-Emitting Diode (LED) light bulb consumes about 7 W (i.e., 7 J/s) and can produce 700 Lumens. The energy efficiency of the light bulb denoted by the produced amount of light over the consumed electrical energy. Table 1.1 compares the energy efficiency of the two light bulbs.

Table 1.1 Comparision between incandescent and LED light bulb

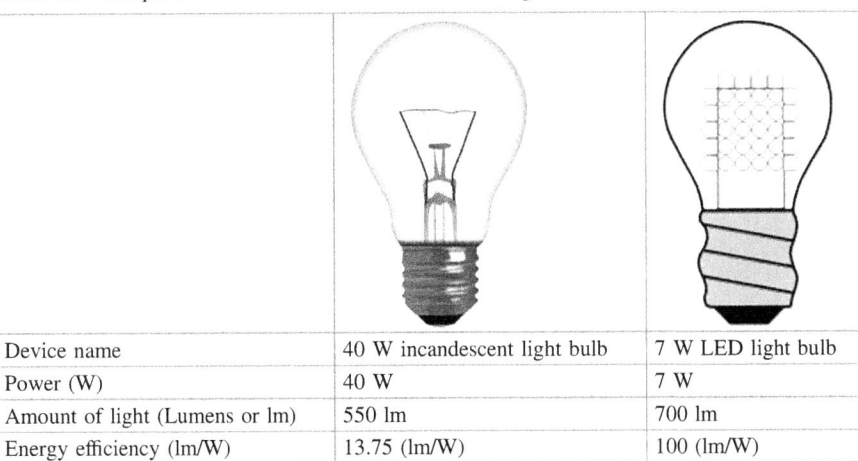

Device name	40 W incandescent light bulb	7 W LED light bulb
Power (W)	40 W	7 W
Amount of light (Lumens or lm)	550 lm	700 lm
Energy efficiency (lm/W)	13.75 (lm/W)	100 (lm/W)

In Table 1.1, we can see that the LED-based light bulb is about seven times more efficient than the traditional incandescent light bulbs. But the question here is: why a 40 W incandescent light bulb is less efficient than 7 W LED light bulb. To answer this question, we need to understand what except visible light to the human eye is produced by these light bulbs. It seems that, except visible light, the light bulb produces heat and invisible light (e.g., Infrared), and the percentage of produced visible light at LED light bulbs is significantly higher compared to incandescent light bulb.

1.2 Power Consumption—Where Does the Power Go?

Complementary Metal-Oxide Semiconductor (CMOS) technology has been a driving force in recent development of computer systems. CMOS has been popular due to its resilience for noise as well as low heat produced during its operation compared to other semiconductor technologies.

CMOS power consumption can be divided into two main categories: **Dynamic power**, and **Static power**. The dynamic power is further divided into switching power and short-circuit power. This section explains what are the parameters that affect power consumption at these categories.

1.2.1 Dynamic Power

Dynamic power consumption is created by circuit activity (i.e., transistor switches, changes of values in registers, etc.) and depends mainly on a specific usage scenario, clock rates, and I/O activity. The sources of the dynamic power consumption are short-circuit current and switched capacitance.

Switching power is the power that dissipates due to switching the transistor from 0 to 1 and vice versa, the dynamic power can be defined as in (1.3).

$$P_{\text{dynamic}} = \alpha \cdot C \cdot V^2 \cdot f, \tag{1.3}$$

where C is the *load capacitance*, V is the *supply voltage*, α is the *activity factor*, and f is the operating *frequency*. The frequency has theoretically linear dependence on the voltage supply, hence from the previous equation, we can see that the dynamic power exhibits a *cubic* dependence on the voltage supply. The peak processor performance is not always necessary, a drop on the operating frequency can be tolerated and this will cubically reduce the dynamic power consumption. Whereas combined reduction of the supply voltage and clock frequency lies in the roots of the widely adopted DPM technique called *Dynamic Voltage and Frequency Scaling* (DVFS). The main idea of this technique is to intentionally downscale the processor performance, when it is not fully utilized, by decreasing the voltage and frequency of the processor that in ideal case should result in cubic reduction of the dynamic power consumption. DVFS is supported by most modern processors including mobile, desktop, and server systems. We will discuss this technique in detail in the following sections.

Short-circuit power is the power dissipated during the brief transitional period when both the n and p transistors of a CMOS gate are "on". Short-circuit current causes about 10–15% of the total power consumption and so far, no way has been found to reduce this value without compromising the performance.

1.2.2 Static Power

Static power is due to the leakage currents on the transistor while they are "off", this power is equal to leakage current (I_{leakage}) times the voltage supply (V), i.e., $I_{\text{leakage}} \cdot V$, where the I_{leakage} can come from several sources, including gate leakage and subthreshold leakage, subthreshold leakage power is given by the following simplified equation:

$$P_{\text{static}} = V \left(ke^{-\frac{qV_{th}}{\alpha K_a T}} \right) \tag{1.4}$$

In this equation, V refers to the *supply voltage*, while V_{th} refers to the *threshold voltage*.[1] The exponential link between leakage power and threshold voltage (V_{th}) is immediately obvious. Lowering the threshold voltage brings a tremendous increase in leakage power. Temperature (T) is also an important factor in the equation: leakage power depends exponentially on temperature. The remaining parameters k, q, α, and K_a, summarize logic design and fabrication characteristics.

Table 1.2 shows the dynamic and static power breakdown at different work points (frequency, voltage) of the Skylake [80] core. The table suggests that when moving from frequency of 1.2–3.5 GHz, both the static and dynamic power grows as the voltage also scales. It should be noted that the static power percentage from the total power is higher at the lower frequency.

1.2.3 Power Savings Techniques

As we saw earlier, the CMOS power consists of dynamic power and static power. Here, we discuss the main techniques used for saving this power components at modern processors. First, we will give an overview of the main parts of the *clocking* and *Power Delivery Networks* (PDN) inside the integrated chips.

1.2.3.1 Clocking

Processors, as many other digital circuits, mostly operate on synchronous digital circuits. Processors typically operate at hundreds of megahertz to multiple gigahertz clock frequencies. The clocks supplied to these processors come from clock generator, Phase Locked Loop (PLL), which multiply a lower frequency reference clock (usually 50 or 100 MHz) up to the operating frequency of the processor. The reference clock is generated using crystal oscillator that is input to the chip. The

Table 1.2 Static and dynamic power for Skylake core at different frequencies and voltages

Frequency (GHz)	3.5	1.2
Voltage (Volt)	1.06	0.64
Temperature (C)	50	50
Static power (Watts)	0.369	0.102
% Static power	8%	16%
Dynamic power (Watts)	4.415	0.551
% Dynamic power	92%	84%
Total power	4.785	0.654

---

[1]The minimum gate-to-source voltage differential that is needed to create a conducting path between the source and drain terminals.

multiplication factor can be quite large in cases where the operating frequency is multiple gigahertz and the reference crystal is just tens or hundreds of megahertz.

As shown in Fig. 1.2, the reference clock enters the chip and drives a PLL, which then drives the system's clock distribution. The clock distribution is usually balanced using, clock grid, clock H-tree, or clock spine so that the clock arrives at every endpoint simultaneously. One of those endpoints is the PLL's feedback input. The function of the PLL is to compare the distributed clock to the incoming reference clock, and vary the phase and frequency of its output until the reference and feedback clocks are phase and frequency matched.

Clock Gating

Clock gating is a power management technique for saving the dynamic power, this technique stops the toggling of the input clock to some component by effectively adding AND gate on the clock with "Clock_Enable" signal as shown in Fig. 1.3.

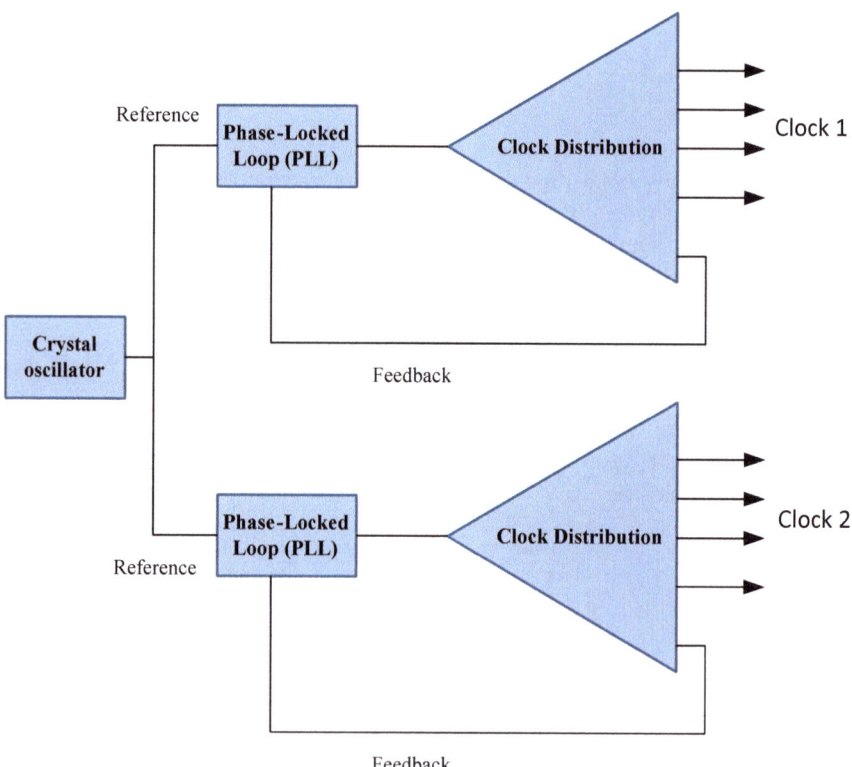

Fig. 1.2 Clock generation and distribution for two clock domains

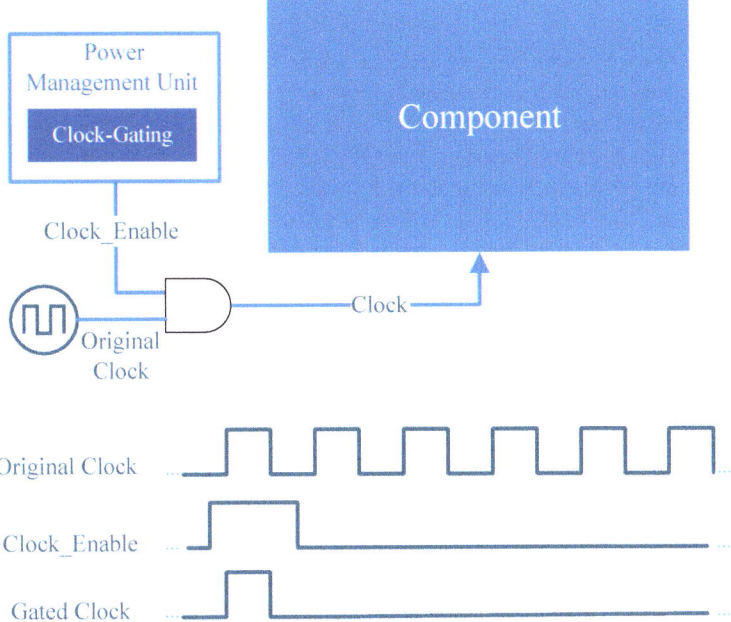

Fig. 1.3 Clock gating of the main clock to some component

The clock gating effectively turns the frequency to be zero (no toggling) or it can be considered as reducing the activity factor down to zero; hence, the resulting dynamic power according to the dynamic power Eq. (1.4) is:

$$P_{\text{dynamic}} = \alpha \cdot C \cdot V^2 \cdot f = P_{\text{dynamic}} = \alpha \cdot C \cdot V^2 \cdot 0 = 0$$

Clock Tree Gating

As previously shown, modern processor has large circuits inside each unit that can work with the same clock domain, in order to reduce the skew (clock signal arrives at different components at different times), clock distribution networks (or clock trees) are used, such as clock grid, clock H-tree, and clock spine. The power used to drive the clock signal can be 10–30% of the total power used by the entire chip. While clock gating ensures that the dynamic power at the component is zero, still the clock tree consumes additional significant power.

One additional technique of saving clocking power is the gating of the clock tree when possible, this is done when all the endpoints of the clock tree are gated or can be gated, a central clock gating on the root of the clock tree or sub-trees can be applied to save the power that is consumed by the clock tree distribution.

PLL Shutdown

Processor has many clock domain and PLLs. The PLL consumes significant amount of power; when the clock tree is gated, the PLL is still consuming power (hundreds of microwatts to few tens of milliwatts) as it is still working. In order to save this power, there is an option to turn off the PLL totally, this can be done once all the clock consumers are clock gated and the system can tolerate the additional latency when the clocks are required and the PLL needs to be turned back on, as turning the PLL ON might take few microseconds in order to completed the PLL locking processes.

Crystal Oscillator Shutdown

The crystal oscillator is normally a small chip on the board that generates the reference clock for PLLs, normally many PLLs are sharing the same reference clock. This small chip consumes power (normally few milliwatts) also when the PLLs connected to it are turned off. For aggressive power saving, it is important to save this power also, this chip can be turned off once all the PLLs using its reference are off and if the latency to turn it back on is acceptable; turning the crystal back on might take tens to hundreds of microseconds.

1.2.3.2 Power Delivery

While techniques such as clock gating are efficient in reducing the dynamic power, the static power is still consumed when only clock gating is applied. As shown previously, the static power is a function of thermal (T) and the voltage (V) beside the design and fabrication characteristics. While, normally we have limited control on the thermals of the system, lowering supply voltage provides an exponential reduction in leakage power.

The PDN at modern processor consists of many components. Figure 1.9 shows an example of two power delivery schemes that are used at modern processors. Main components of power delivery scheme are shown at the following subsections.

Battery

Batteries are the main source of electrical energy in mobile devices. The energy of chemical compounds acts as storage medium, and during discharge, a chemical process occurs that generates energy which can be drawn from the battery in form of an electric current at a certain voltage level.

For a number of batteries, this process can be reversed and the battery recharged, i.e., the intake of electric energy can restore the chemical composition that contains higher energy and can closely reestablish the original structures within the battery.

Lithium-ion batteries are widely used in mobile devices such as laptops, with some thinner models using the flatter lithium polymer technology. These two technologies have largely replaced the older nickel metal-hydride batteries.

Battery life is highly variable by model and workload, the battery life metric is watt-hours (Wh), that represents the amount of energy stored inside the battery, for example:

- Battery with 20 Wh can last about 10 h if the average power of the workload is 2 W.

Another metric which is widely used is the amp-hours (Ah) or milliamp (mAh) which represent the amount of current that can be drawn per hour from the battery at a given voltage. This metric can be easily converted to Wh by simply multiplying the number of amp-hours by the output voltage of the battery, for example:

- Given a 3 Ah (3000 mAh) battery with output voltage of 5 V, then energy is 3 Ah · 5 V = 15 Wh

Battery's performance gradually decreases over time; substantial reduction in capacity is typically evident after 1 to 3 years of regular use, depending on the charging and discharging pattern and the design of the battery.

Power Supply Unit (PSU)

A Power Supply Unit (PSU) is an instrument that converts alternating current (AC), that comes from wall socket for example, to low Direct-Current (DC) voltage. PSU supplies multiple power supply rails to the computer system with multiple voltage levels.

Most modern desktop personal computer PSUs are conforming to the Advanced Technology eXtended (ATX) specification, ATX is a motherboard configuration specification developed by Intel in 1995, which defines form factor and voltage tolerances requirements. For example, while an ATX power supply is connected to the mains supply, it always provides a 5 V standby (5 V_{SB}) voltage so that the standby functions on the computer and certain peripherals are powered. ATX power supplies are turned on and off by a signal from the motherboard (PS_ON). They also provide a signal (PWR_OK) to the motherboard to indicate when the DC voltages are stabilized and within specification, so that the computer is able to safely power up and boot.

Voltage Regulators

A voltage regulator (VR) is an electronic circuit that provides a stable DC voltage independent of the load current, temperature, and AC line voltage variations. There are different types of voltage regulators, such as Linear Voltage Regulators (LVRs) and switching regulators.

LVR works as variable resistance, where the resistance of the regulator varies in accordance with the load resulting in a constant output voltage. As shown in Fig. 1.4, the LVR continuously adjusts a voltage divider network to maintain a constant output voltage and continually dissipating the difference between the input and regulated voltages as waste heat.

The major drawback of using linear regulators can be the excessive power dissipation of its series transistor operating in a linear mode. As explained previously, a linear regulator transistor is conceptually a variable resistor. Since all the load current must pass through the series transistor, its power dissipation is $P_{Loss} = (V_{IN} - V_O) \cdot I_O$. In this case, the efficiency of a LVR (η_{LVR}) can be estimated by

$$\eta_{LVR} = \frac{P_{Output}}{P_{Input}} = \frac{P_{Output}}{P_{Output} + P_{Loss}} = \frac{V_O \cdot I_O}{V_O \cdot I_O + (V_{IN} - V_O) \cdot I_O} = \frac{V_O}{V_{IN}} \quad (1.5)$$

The LVR efficiency shown in Eq. (1.5) is the upper bound, normally additional losses exist in the circuit which reduces the efficiency. The upper bound on the LVR efficiency as a function of the V_O/V_{IN} ratio is plotted in Fig. 1.5.

While for high difference between V_{IN} and V_O the LVR efficiency might be very low, the LVR can be very efficient if the V_O is very close to the V_{IN}. LVRs that can work at low headroom (i.e., $V_{IN} - V_O$) are called Low-Dropout Regulators (LDOs). Designers use LVR or LDOs because they are simple, low noise, low cost, easy to use, and provide fast transient response, while in case of V_{IN} close to V_O, an LDO may have high efficiency.

Switching Mode Power Supply (SMPS)—also called switching regulator—is a voltage regulator that uses a switching element to transform the incoming power supply into a pulsed voltage, which is then smoothed using capacitors, inductors, and other elements as shown in Fig. 1.6. Power is supplied from the input to the output by switching ON a switch (MOSFET) until the desired voltage is reached. Once the output voltage reaches the target value, the switch element is switched to OFF and no input power is consumed, this is why it is called switching regulators. Repeating this operation at high speeds makes it possible to supply voltage efficiently and with less heat generation.

The main reasons that designers use the SMPS are: High efficiency, low-power dissipation, and small size. While LVRs and LDOs are simpler, in high-current applications, it is more efficient to use SMPS. For example, a 12 V input, 5 V output SMPS can usually achieve >90% efficiency versus less than 41.6% for a LVR.

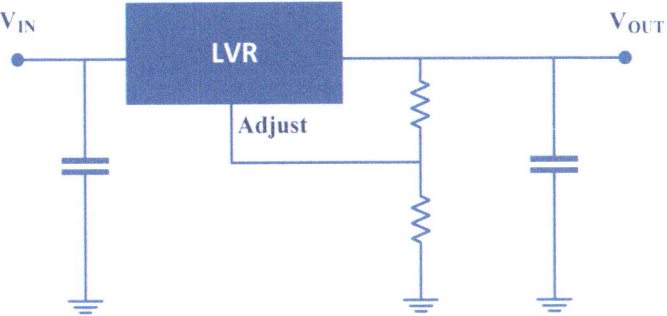

Fig. 1.4 LVR voltage regulator scheme

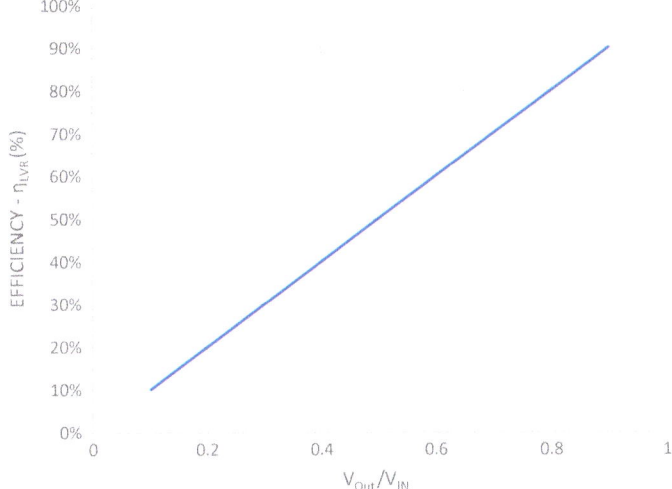

Fig. 1.5 The upper bound on the LVR efficiency (η_{LVR}) versus V_O/V_{IN} ratio

Fig. 1.6 Buck converter
SMPS simplified scheme

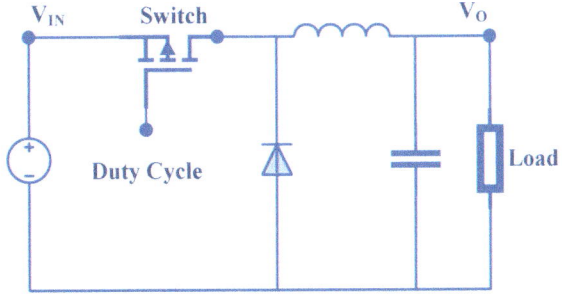

The main three types of SMPS are as follows:

- **Boost converters**—also called step-up converter—is a DC-to-DC power converter that steps up voltage from its input voltage (V_{IN}) to its output voltage (V_O).
- **Buck converters**—also called step-down converter—is a DC-to-DC power converter which steps down voltage from its input voltage (V_{IN}) to its output voltage (V_O).
- **Buck–boost converters**—this type of converter is a DC-to-DC converter that has an output voltage magnitude that is either greater than or less than the input voltage magnitude.

Modern processors use mainly the buck converter of the SMPS, they convert the voltage from (7.2–20 V) to less than 2 V for the processor's logic power rails and higher voltage for the IOs (e.g., 3.3 and 5 V).

The nonidealities of the power devices, such as switches, diodes, and inductors, account for the bulk of the power losses in the converter, while the power losses due to the control circuitry are usually insignificant. Both static and dynamic power losses occur in any switching regulator. Static power losses include I^2R (conduction) losses in the wires or board traces, as well as in the switches and inductor, as in any electrical circuit. Dynamic power losses occur as a result of switching, such as the charging and discharging of the switch gate, and are proportional to the switching frequency. Normally the efficiency of SMPS is high, however, it is not constant, it is rather a function of the load (P_{out}) and the input voltage (V_{in}) as shown in Fig. 1.7.

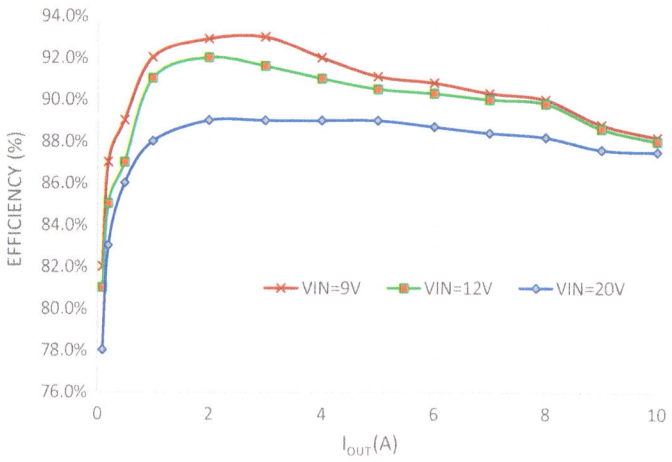

Fig. 1.7 Example of buck converter efficiency curves as a function of I_{OUT} and the input voltage (V_{IN})

Power Gates

A common technique used in integrated circuits to save static power is the Power Gating (PG). Power gating cuts off the voltage to some component by adding high-threshold voltage (V_t) sleep transistors at the input voltage into the circuit as shown in Fig. 1.8. Power gating has many parameters to consider. The Power Gate size must be selected big enough to handle the amount of switching current at any given time. In addition, gate control slew rate is an important parameter that determines the power-gating efficiency. When the slew rate is large, it takes more time to switch off and switch on the circuit and hence can affect the power-gating efficiency. Slew rate is controlled through buffering the gate control signal. Additional parameter to take into account is the simultaneous switching capacitance, the amount of circuit that can be switched simultaneously without affecting the power network integrity, as if a large amount of the circuit is switched simultaneously, the resulting rush current can compromise the power network integrity. The Power Gates circuit needs to be switched in stages in order to prevent this. While Power Gates reduce the static power of the gated components, Power Gates are made of active transistors that themselfs have leakage, hence, leakage reduction for these transistors is an important consideration to maximize power savings.

It is important to notice that when power-gating a circuit, then the state of the circuit (at memory elements) will be lost. Hence normally, power-gating sequence

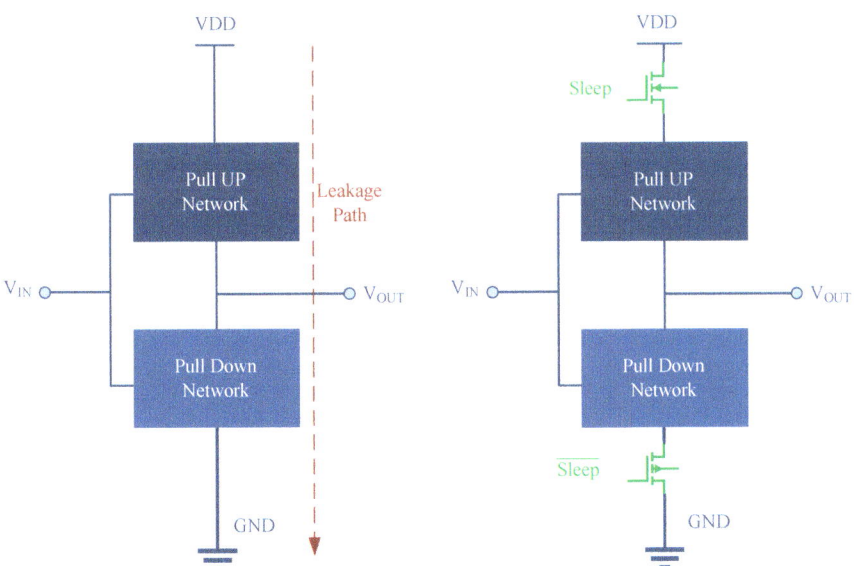

Fig. 1.8 Power-Gating concept: On the left—leakage path exists from the power supply (VDD) to the ground (GND). On the right—adding sleep transistor (high-threshold PMOS and NMOS), when Sleep = 0 the circuit in powered ON, when the Sleep = 1 the circuit is OFF

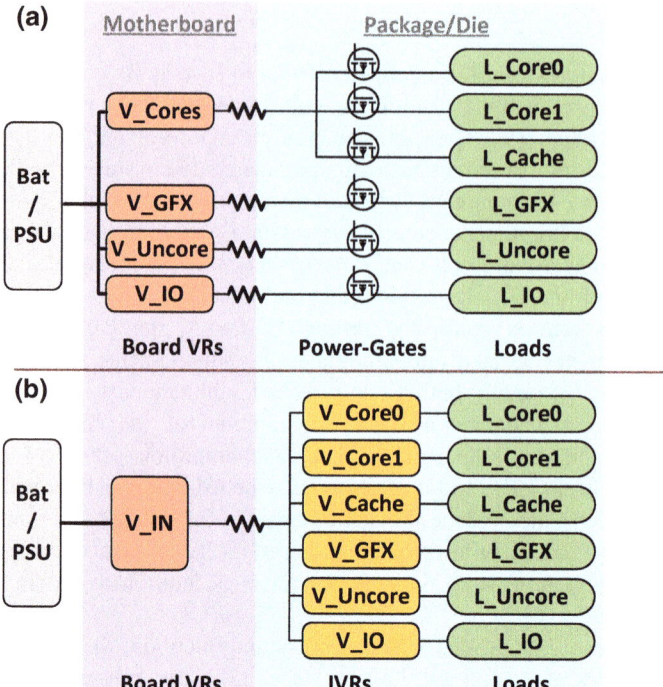

Fig. 1.9 Example of Power Delivery Network (PDN) of modern processor. **a** shows a PDN using motherboard voltage regulators (**b**) shows a PDN that uses Integrated Voltage Regulators (IVR)

includes a flow for saving the state before power-gating the circuit (asserting Sleep in Fig. 1.8), and restoring the state after exiting power-gating mode of the circuit before invoking the power gating (de-asserting Sleep). There are two main methods to retain the state when power-gating:

- Always-on Memory—In this method, before power gating, the state of the circuit is read and saved into a memory (on-chip or off-chip) where it's power supply is kept ON after activating the Power Gates, normally, an always-on power domain is chosen for the state save memory.
- Retention Flops—in some cases, retention flops/registers are used to store the state of the gated circuit. Retention registers are special low leakage flip-flops used to hold the data of the main registers of the power-gated block, this is done by special designed flip-flops that have *dual supply* (normally gated and un-gated domain of the same supply). Thus, the internal state of the block during power down mode can be retained in place and loaded back to it when the circuit is reactivated. Retention registers are always powered up, the retention strategy is design dependent, a power-gating controller controls the retention mechanism and when to save the state of the power-gated circuit and when to restore it back.

Retention Voltage

Once a circuit is idle, it can be clock gated in order to save the dynamic power. Reducing the static power requires reducing the voltage or/and temperature. While reducing the voltage to zero (e.g., using Power Gates) saves all the static power, this operation requires preserving the state of the registers before cutting off the voltage and restoring them back at wake-up (returning the voltage), this technique requires special design and flow, in addition, long latency for enter and exit the state.

Additional option—besides reducing the voltage to zero—is to reduce the voltage to Data Retention Voltage (DRV), a voltage level below the minimal operational voltage, this voltage reduction can reduce the leakage power exponentially significantly while preserving the state of the circuit. DVR is the voltage below which a bit cell (flip-flop/latch/SRAM) loses its data.

This mechanism can be applied when the designer has a control on the voltage level of the controlled component, this is normally done by directly controlling the voltage regulator output level, which supplies power to the component.

Power Delivery Network

Power delivery network of modern processors uses all previous mentioned components and techniques, Fig. 1.9 shows two power schemes that are used in modern processor:

- **Power delivery scheme (a)**:

  - The scheme shows six loads (two cores, cache, graphics, Uncore, and IOs)
  - Four motherboard voltage regulators (V_Cores, V_GFX, V_Uncore and V_IO)
  - Three on-die Power Gates for the Core0/1 and Uncore domains, in addition to Power Gate per each one of the other domains.
  - The cores and cache are sharing the same voltage regulator (V_Cores), this means that voltage level at all domain is the same when all three domains are active.

This scheme is intended for a high-end processors' System on Chip (SoC) with motherboard voltage regulators, such as Intel's® 2nd, 3rd, 6th, and 7th Generation Core™ microprocessors [49–52].

- **Power delivery scheme (b)**:

  - Have the same six loads from the above scheme
  - One motherboard voltage-regulator (V_IN)
  - Six different on-die/package Integrated Voltage Regulators (IVRs)

- The IVR is used as Power Gates also when the voltage needs to be turned off to some domain.
- This scheme enables having different voltage levels for each one of the cores and the caches.

This scheme is typical for high-end processors' SoCs such as the Intel's® 4th and 5th Generation Core™ microprocessors [53, 54].

1.3 Power Management Methods

Many research works have been done in the area of power management and energy efficiency in computing systems. As power and energy management techniques are closely connected, literature mostly refer to them as *Power Management* (PM). As shown in Fig. 1.10, from the high-level, power management techniques can be divided into SPM and DPM.

1.3.1 Static Power Management

1.3.1.1 Hardware Level

From the hardware point of view, SPM contains all the optimization methods that are applied at the design time at the circuit, logic, architectural, and system levels [21].

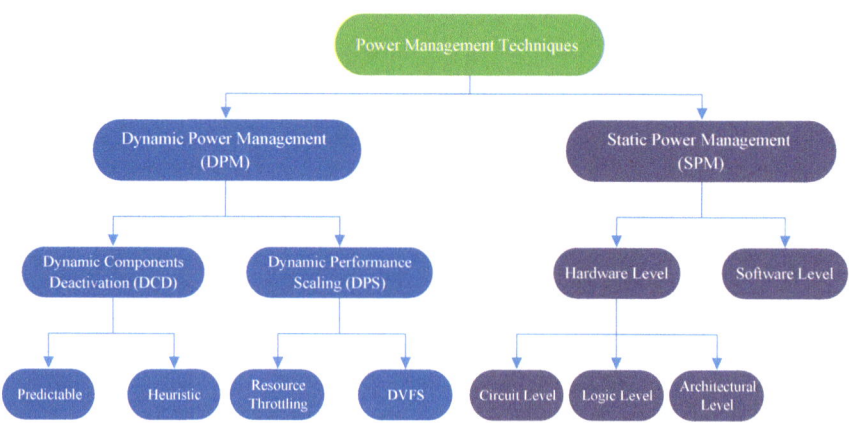

Fig. 1.10 High-level power management techniques

Circuit Level

Circuit-level optimizations are focused on the reduction of switching activity power of individual logic gates and transistor level combinational circuits by the application of a complex gate design and transistor sizing.

Logic Level

Optimizations at the logic level are aimed at the switching activity power of logic level combinational and sequential circuits.

Architecture Level

Architecture level methods include the analysis of the system design and subsequent incorporation of power optimization techniques in it. This kind of optimization refers to the process of efficient mapping of a high-level problem specification onto a design.

1.3.1.2 Software Level

Beside hardware level optimizations, it is very important to optimize the software that will be running on top of the hardware. Poor software design can dramatically affect the performance and the power of the system, thus optimizing the hardware itself is not sufficient. For example, implementing a simple sorting problem of integer numbers using different algorithms results in different energy consumption and performance. Implementing the problem with *bubble sort* will result in reduced performance and higher energy consumption relative to more optimized algorithm such as quick-sort when running on the same system.

In addition to the selection of proper algorithm, the use of an optimized compilation process and flags might dramatically affect energy consumption, for example, at a system that supports Single-Instruction Multiple-Data (SIMD) instruction, such as Intel AVX [24], using modern compilers that can utilize these instructions by applying auto-vectorization techniques or by choosing software libraries optimized for vector instruction; such as Intel Math Kernel Library(MKL) [23]; will result in more energy-efficient program.

1.3.2 Dynamic Power Management

DPM techniques include methods and strategies for runtime adaptation of a system's behavior according to current resource requirements or any other dynamic characteristic of the system's state. At modern high-performance processors, these techniques try to maximize the performance and energy efficiency of the system according the user or operating system predefined policy. The optimization space has several constraints that need to be taken into account by the Power Management Unit (PMU) of modern processors, this includes thermal and power delivery constraints at multiple levels of the system.

DPM utilizes the fact that most systems experience variable workloads during the operation time allowing the dynamic adjustment of power states according to current performance requirements. DPM techniques can be classified as Dynamic Performance Scaling (DPS), such as DVFS, and partial or complete Dynamic Component Deactivation (DCD) techniques such as clock gating of parts of an electronic component or complete disabling during periods of inactivity. DPM techniques are mostly implemented at hardware and firmware levels, while software DPM techniques utilize interface to the system's power management by applying hardware implementations of the DPM according to policies set by the software (mostly operating system).

Appling optimal DPM is a difficult problem to solve, modem processors use heuristic algorithms to solve this problem. The problem could be easily solved if transitions between power states would cause negligible energy and performance overhead. However, when applying DPM techniques, the processor enter low-power states, which lead to additional energy consumption and delays, caused by the re-initialization of the components when going back to higher power state. For example, if entering a low-power state requires shutdown of the power supply, returning to the active state will cause a delay consisting of: turning on and stabilizing the power supply, Power Gates, and clocks, re-initialization of the system, and restoring the context. In the case of non-negligible transitions, efficient power management turns into a difficult optimization problem. A transition to low-power state is worthwhile only if the period of inactivity is longer than the aggregated delay of transitions from and into the active state, and saved energy is higher than required to reinitialize the component.

1.3.2.1 Dynamic Component Deactivation (DCD)

In modern processors, DCD, also known as idle power management, is a technique for reducing the power consumption of components when they are inactive. In principle, components at the processor that are inactive can be turned off to save energy and power. The problem is trivial in the case of a negligible transition overhead. However, in reality such transitions lead not only to delays, which can degrade performance of the system, but also to additional energy draw. Therefore, to achieve efficiency, a transition has to be done only if the idle period is long enough to cover the transition overhead. Hence, modern processors define multiple power states, that trade-off transition's latency with power level, these states are applied at different idleness periods of the workload. Figure 1.11 shows example of different idle power states of a processor with the corresponding enter/exit latency difference and power levels.

In most real-world systems, there is a limited or no knowledge about the future workload. Therefore, a prediction of an effective transition has to be done according to historical data or some system model. A large volume of research has been done to develop efficient methods to solve this problem [25, 26]. DCD techniques can be divided into predictable and heuristic-based techniques as shown in Fig. 1.10.

Predictable

The behavior of part of the processor's components can be predicted, where the PMUs that manage the component have enough knowledge when the component can be deactivated and when it will need to be moved back to active state. Accurate indicator of such behavior comes from operating system, drivers or Input–Output (IO) channels. For example, a camera accelerator can be turned off all the time when the camera is not in use, the indication that the camera is not in use can come from the camera driver. Other example is the display controller, which reads frames from the main memory and sends the frame's data to the external panel, this component normally has internal buffer that stores part of fame's data, a watermark of this buffer can be used as indication that the buffer is getting empty and needs to be refilled, at the time that the watermark is not reached (buffer is full enough), the subcomponent that reads data from memory can be turned off (e.g., clock gated) assuming that no other component in the processor needs this subcomponent.

Components that have deterministic behavior based on inputs and indicators from the system can be power-managed easily and in most cases, all the power saving opportunities can be fulfilled.

Heuristic

Heuristic-based techniques attempt to correlate between the past history of the system behavior and its near future for the managed component. The efficiency of such techniques is highly dependent on the actual correlation between past and future events and quality of tuning for a particular type of the workload. A nonideal prediction can result in an overprediction or underprediction. An overprediction means that the actual idle period of the component is shorter than the predicted leading to a performance and energy penalty. On the other hand, an underprediction means that the actual idle period is longer the predicted. The latter case does not have any influence on the performance; however, it results in reduced energy savings.

Some heuristic techniques utilize threshold for a real-time execution parameter to make predictions of idle periods. The idea is to define the length of time threshold after which a period of idleness of the component can be treated as long

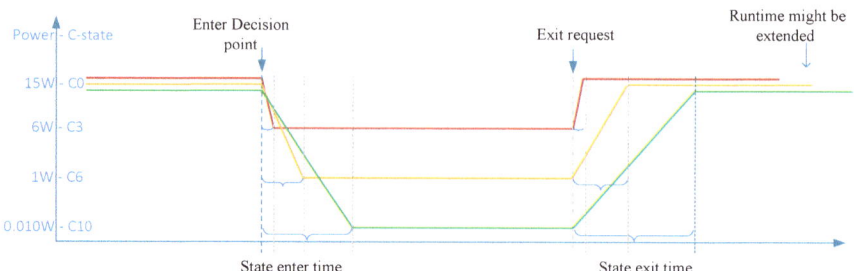

Fig. 1.11 Idle states (e.g., C-states at Intel architecture) enter and exit. Deeper idle state has reduced power level while latency to enter and exit from state is higher than the shallower idle state. Enter deep state might affect performance (e.g., system responsiveness) as exit time is relatively high

enough to do a transition to a low-power state (e.g., turn off clocks, power gating, etc.). The activation of the component is initiated once the first request to a component is received. One of the advantages of this policy is that it can be applied to any workload type, while the overpredictions and the underpredictions can be controlled by adjusting the value of the timeout threshold. This type of heuristic is more suitable for components that their workload is deterministic as the policy requires adjustment of the threshold value for each workload if the workloads are variable. Applying this kind of heuristics to changing workload can lead to a performance loss on the activation, or losing of energy savings opportunities.

To overcome the drawbacks of the fixed timeout policy, some more advanced heuristics are used, these heuristics attempt to predict the idle and active times of the component based on previous periods of inactivity while assuming that they are highly correlated with the nearest future. According to the analysis of the historical information, they predict the length of the next idle period before it actually starts. These policies are highly dependent on the actual workload and strength of the correlation between past and future events. Predictive wake-up techniques aim to eliminate the performance penalty on the activation. The transition to the active state is predicted based on the past history and performed before an actual user request (event) by applying pre-wake-up actions (e.g., turn-on power supply, restore clocks) as shown in Fig. 1.12. These techniques may increase the energy consumption, but reduce performance losses on wake-ups.

Modern processor uses several heuristics-based algorithms for power managements, these heuristics attempt to forecast the idle duration and next interrupt for the managed component by looking at the history and sometime attempt to corollate the future prediction to indicators from the system resources, such as number of active cores, display on/off indication, PCIe activity, camera on/off indication and performance counters, all these indicators fed into the decision algorithm that is normally implemented in the power management hardware or firmware. We give examples of such algorithms in the coming sections.

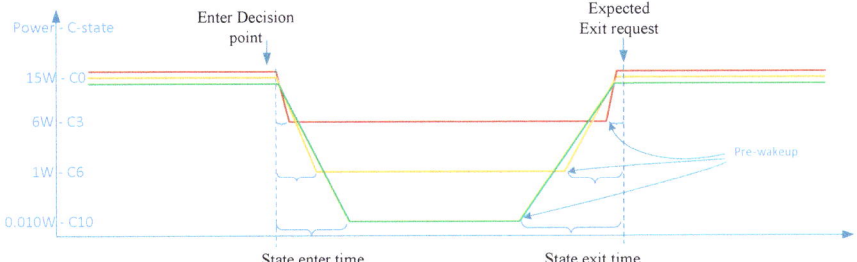

Fig. 1.12 Pre-wake-up—In order to reduce the performance and energy effect of exit time from idle state, a pre-wake-up is used in case that the exit time is predictable (e.g., timer-based exit)

1.3.2.2   Dynamic Performance Scaling (DPS)

DPS includes different techniques that can be applied to computer components supporting dynamic adjustment of their performance proportionally to the power consumption. Instead of complete deactivations, some components, such as processor's cores, allow gradual reductions or increases of the clock frequency along with the adjustment of the supply voltage in cases when the resource is not utilized for the full capacity. This idea lies in the roots of the widely adopted DVFS technique. Some resources do not allow or have the capability of changing the frequency and voltage of the component, in this case other techniques are used for limiting the performance, such as duty cycling of the component, this technique enables and disables the component in a periodic manner to reduce the performance of the component. Duty cycling technique is effective in reducing the performance, it is not energy efficient compared to DVFS.

Dynamic Voltage and Frequency Scaling (DVFS)

DVFS exploits the fact that different performance levels are required during the execution of computing workloads, a trade-off between power and performance is done in order to raise the energy efficiency, for example, running web browser or video playback normally requires less processors performance than workloads that does many arithmetic calculations like MATLAB or Excel programs. At lighter workloads, the processor should be running at reduced frequency relative to the more performance intensive workloads.

Although the frequency can be adjusted separately, frequency scaling reduces the dynamic power only and the reduction is by linear factor. Saving exponential factor of the static power and quadratic factor in the dynamic power requires dynamic voltage scaling too. Processors are normally running at the maximum possible frequency of a given voltage, the graph that describes the minimum required voltage to run at certain frequency is called the voltage/frequency (V/F) curve as shown in Fig. 1.13.

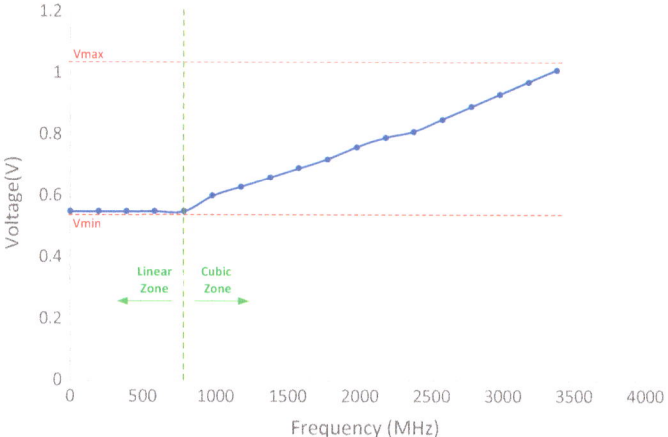

Fig. 1.13 Theoretical voltage/frequency curve for a processor's core. V_{min} denotes the minimal functional voltage while V_{max} denotes the maximum allowed voltage

There are two main zones at the V/F curve: the *Linear zone* and the *Cubic zone*.

- **Linear zone**—describes the behavior of the voltage and frequency near the minimum functional voltage (V_{min}), at this zone the voltage is not scaling below V_{min}, while different frequencies are allowed to scale down, starting from 0 MHz frequency until the maximum possible frequency at this voltage—in the example shown in Fig. 1.13, this frequency is around 800 MHz. The power consumption at this linear zone is linearly correlated to the frequency as shown in Fig. 1.14.
- **Cubic zone**—at this zone, the power is cubically proportional to frequency as shown below:

  - Theoretically, the frequency (F) and voltage (V) relation can be described as

$$ f \approx k \cdot V $$

  - In addition, the dynamic power can be described as,

$$ P_{dynamic} = \alpha \cdot C \cdot V^2 f \approx \alpha \cdot C \cdot (k \cdot f)^2 f \approx k \cdot \alpha \cdot C \cdot f^3 $$

$$ P_{total} = P_{dynamic} + P_{static} \propto f^3 $$

In parctice, at the cubic zone, the voltage required to run a circuit tends to increase with the square of the frequency; the exact relationship between these two components is based on the transistor design. Different transistor designs and process technologies have different characteristics. Transistors that can achieve higher frequencies must trade off low-power characteristics. These are commonly

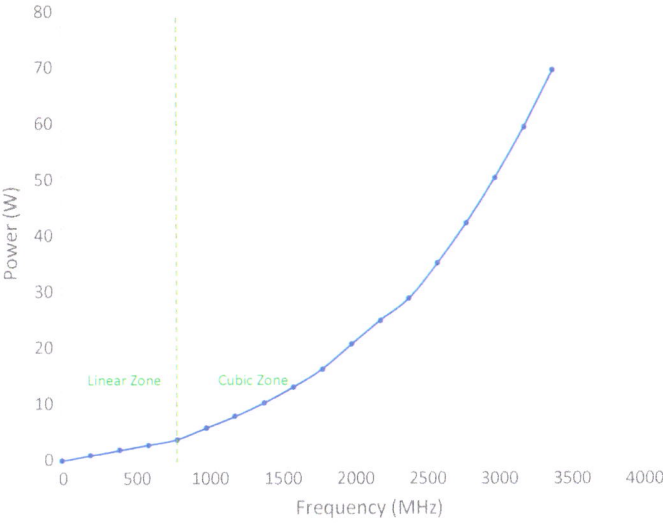

Fig. 1.14 Power and frequency relationship at linear and cubic zones

used in high-power server processors designs. On the other hand, transistors can be optimized for low leakage and low-power operation, trading off high frequency operation. This type of transistor is used in devices such as: phone, tablet, and laptop. It should be noted that the efficiency of some electrical components, such as voltage regulators, decreases with a temperature increase, moreover, the static power consumption increases with temperature (see static power formula (1.4)), on the other hand increasing power consumption raises the temperature, hence, increases in voltage or frequency may raise the system power demand even faster than the CMOS formula indicates, and vice versa.

DVFS can provide substantial energy savings; however, it has to be applied carefully, as the result may significantly vary for different hardware and software system architectures. DVFS reduces the number of instructions a processor can issue in a given amount of time—Instructions Per Second (IPS), thus reduces the performance. This, in turn, increases runtime for program segments, especially programs that are CPU-bound. Hence, it is a challenging task to develop algorithm for controlling the DVFS. Although the application of DVFS may seem to be straightforward, real-world systems raise many complexities that have to be considered as listed here:

- Due to complex architectures of modern processors (i.e., pipelining, multilevel cache, etc.), the prediction of the required CPU clock frequency that will meet application's performance requirements is not trivial. Some algorithms try to figure this dynamically with searching algorithm for finding the optimal frequency for the current phase of the program.

- In contrast to the theory, power consumption by a processor may not be quadratic to its supply voltage. For example, in [27], it is shown that some architectures may include several supply voltages that power different parts of the chip, and even if one of them can be reduced, overall power consumption will be dominated by the larger supply voltage. Energy efficiency point might not be the one with the minimal CPU core power that still satisfies the performance requirement, this point should be considered for the whole SoC and not just the core that executes the program instructions, it might be more efficient to run the core at higher frequency, finish the work early, and let the system go to sleep early [29], this technique is called Race to Halt (R2H).
- Moreover, execution time of the program running on the processor may not be inversely proportional to the clock frequency, and DVFS may result in non-linearities in the execution time [28]. For example, if the program has low scalability factor (memory or I/O bounded), CPU core speed will not have a dramatic effect on the execution time.
- Slowing down the CPU core may lead to changes in the order in which tasks are scheduled [27].

DVFS algorithms—approaches that apply DVFS to reduce energy consumption by a system can be divided into three categories:

- *Interval-based*—Interval-based algorithms utilize knowledge of the past periods of the CPU activity [30, 31]. Depending on the utilization of the CPU during previous intervals, they predict the utilization in the near future and appropriately adjust the voltage and frequency. This method is used by part of the operating systems (OS). OS implements an *idle task* that is loaded by the OS scheduler when no other task is available and ready to run. If Idle_Time is the time that is spent at idle task inside a given time frame (Frame_Time) and Busy_Time is the rest of the time at this frame, then CPU utilization (U) can be estimated as follows:

$$U = \frac{Busy\_Time}{Busy\_Time + Idle\_Time} = 1 - \frac{Idle\_Time}{Busy\_Time + Idle\_Time} = 1 - \frac{Idle\_Time}{Frame\_Time}$$

While this method is applied at many operating systems, it is still a heuristic, and not the optimal algorithm. Wierman et al. [32] and Andrew et al. [33] have conducted analytical studies of frequency scaling algorithms in processor systems. They have proved that no online energy-proportional frequency scaling algorithm can be better than the offline optimal algorithm.

- *Inter-task*—at these approaches, instead of relying on coarse-grained information on the CPU utilization, it distinguishes different tasks running in the system and assign them different speeds [34, 35]. In general, the problem is not easy, as the workloads are not regular and do not have constant behavior over time, some of the workloads change behavior over time with different inputs or system status.

- *Intra-task*—In contrast to inter-task, intra-task approaches use fine-grained information about the structure of programs and adjust the processor frequency and voltage within the tasks [36, 37]. These algorithms analyze the program and attempt to figure its phases. Such policies can be implemented by splitting a program execution into phases and assigning different CPU frequencies to each of them. Another way is to implement them at the compiler level. This kind of approaches utilizes compiler's knowledge of a program's structure to make inferences about possible periods for the clock frequency reduction. Although this strategy outperforms the other two at most cases, it requires program analysis (offline or online) which might be overhead and not practical for real-world applications.

Race to Halt (R2H)—Energy efficiency is an important factor in modern processor. DVFS is a common mechanism for achieving best performance at a given power budget by controlling the voltage and frequency of the processor's cores. Various algorithms exist for DVFS controlling, e.g., demand-based algorithms [42]. However, controlling the core frequency and voltage has limited impact on the overall platform energy efficiency, due to the energy consumption of the other platform components. While lowering the frequency and voltage of the core reduces the energy consumption of the core and might put the core at the most energy-efficient work point, the computing time is lengthened resulting in an increase in the energy consumption of other platform components.

Figure 1.15 shows a conceptual energy consumption as a function of core frequency, two ranges are shown, left-side range is a frequency range where the target performance that is required by the running workload is not met and the right side shows the range of frequency where target performance can be achieved. *F1* frequency point has the minimal core energy and is the energy efficiency frequency if we consider the core only power, while in real life, we should consider the platform energy as this is the energy consumed at the end from the power source (battery/wall socket). Considering the platform energy, it can be seen that the actual platform minimal energy is at *F2* point which is a higher frequency point than F1. R2H algorithms attempt to achieve minimal platform energy by running at higher frequency than the minimal frequency required to meet target performance (F1) in order to halt other platform components and allow them go to deep idle sleep state early, therefore, F2 point can be seen as the balance point that guarantees minimal platform energy, where, in one hand the core consumes higher energy relative to F1, and on the other hand the other platform components are consuming less energy as they are going to sleep states earlier compared to F1.

Many algorithms [43, 44] were developed in order to determine dynamically at runtime the most efficient frequency.

Memory DVFS—memory consumes significant portion of the platform power for client and servers' segments, in client, the power consumption can be 5–10% (depends on the workload scenario) [47] of the system power, while in servers, the power can even reach the 40% [48]. Moreover, additional power inside the processor is consumed at memory controller and memory IO that are part of the memory subsystem.

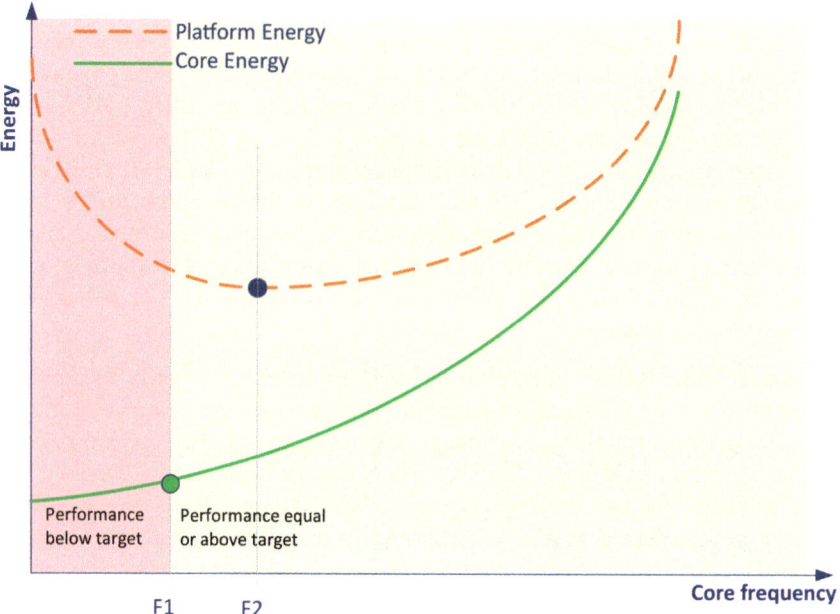

Fig. 1.15 Core and platform energy as a function of core frequency. At F1 core frequency (minimal frequency that meets target performance), the core energy is minimal, while platform energy is minimal at F2 core frequency

Modern processors implemented features for dynamically changing the voltage and frequency of the memory subsystems, this includes frequency of the external memory (DRAM), frequency of the memory controller, and frequency of the fabric that connects the processor components to the processor.

Heuristics inside the processor are used in order to control the memory DVFS, these heuristics attempt to predict the required memory bandwidth and adjust the memory frequency and voltage accordingly, various hints can be used for this prediction, for example: number of active displays, number of active cores, core and graphics frequency, camera enabled/disabled, etc.

A control unit, normally part of the global PMU of the CPU, controls the memory voltage and frequency of the memory. Processor architects build a separate clock and voltage domains for the controlled parts, this includes fabrics, memory controller, DDR IO, and the external memory as shown in Fig. 1.16.

Resource Throttling

Beside the DVFS technique for controlling the power and performance of the processor or its components, other methods exist, such as *resource throttling*. Resource throttling is normally used at components that do not support DVFS, for example, if a component "A" performance is allowed to be reduced at some point in time but its clock is shared with other component "B" that still needs the higher

Fig. 1.16 Memory DVFS scheme. Events are captured in the Memory DVFS control unit, fed into algorithm for controlling the memory subsystem frequencies and voltages. Separate and adjustable voltage and frequency domain are required

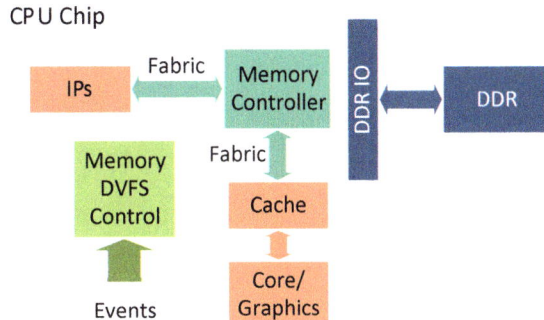

frequency then it is possible to apply resource throttling for component "A" in order to limit its performance.

Other cases were resource throttling may be used, is when a fast response is required, for example, if there is an urgent thermal event that needs fast response in order to reduce the power, then using DVFS might be slow (may take up to few microseconds) and a faster mechanism would be resource throttling. Here, we will see examples of resources throttling.

CPU throttling—Figure 1.17 describes a throttling for CPU core, the component is normally fed with the "Original Clock", when integrating a throttling mechanism, a control is added to the clock, this control uses run-hold (enable/disable) duty cycling to enable or disable the clock toggling at some cycles, normally the modulation ratio is controlled by Configuration Register (CR), while a Finite-State Machine (FSM) controls the clock enable based on the CR value.

While the above mechanism effectively reduces the clock frequency, still we have higher power losses compared to DVFS, for example, the clock grid that is connecting the component to the Phase-Locked Loop (PLL) will still be running at the original higher frequency, in addition, the voltage is at high level which is corresponding to the original higher frequency, hence, only linear part (frequency) of the dynamic power is saved while the static power remains the same as before throttling.

Memory throttling—Memory throttling is a power management technique that is currently available in many commercial systems. Memory throttling restricts read and write traffic to main memory as a means of controlling power consumption. A significant fraction of memory power is proportional to read and write bandwidth, and restricting bandwidth creates an upper bound for memory power. Altering memory throttling dynamically tailors the access rate to workload demands and variable power allocation, this mechanism is useful for regulating external memory temperature [38].

Similar to clock CPU throttling, that limits core cycles within a time frame, memory throttling regulates read and write accesses within a time frame. There are several implementation approaches to memory throttling features. One technique is similar to clock throttling with run-hold duty cycles (enable/disable), where memory accesses pass through at the requested rate untouched during the run portion of a time interval, then are halted during the hold portion [39]. Another

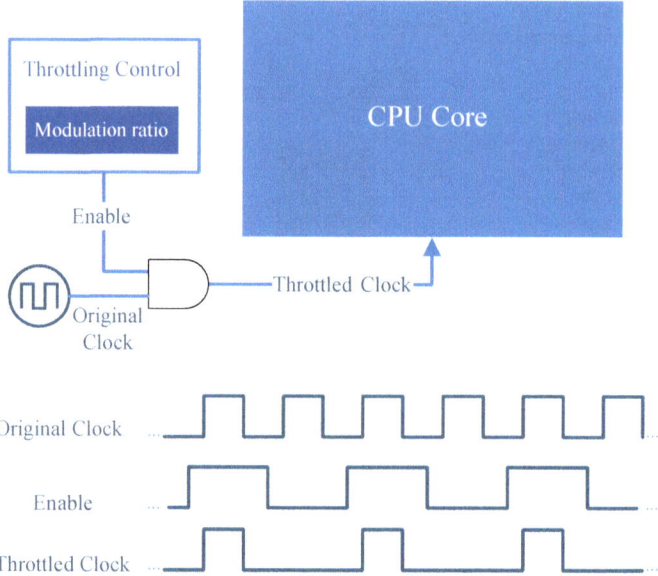

Fig. 1.17 Component throttling with forcing clock modulation using enable signal on the clock

memory throttling mechanism is to periodically insert one or more idle clock cycles between memory accesses [38], spreading out the accesses and lowering peak bandwidth of the memory. Additional mechanism is a memory throttling that allows memory accesses (read/write) within a time interval up to a quota as shown in Fig. 1.18.

Cache resizing—On-chip caches consume larger area and power, while these caches, in many cases, raise the chip performance and reduce its average power, there are workloads that do not benefit from large caches (such as core bounded workloads). Systems architects use smart techniques to attack the high static and dynamic power consumed by on-chip caches. Modern processors use many techniques to make the caches more energy efficient, these techniques power down cache components, such as cache lines, sets, or ways (shut them off or maintain them in low-voltage state) based on algorithms predicts when it is energy efficient to shrink the cache [40, 41] (Fig. 1.19).

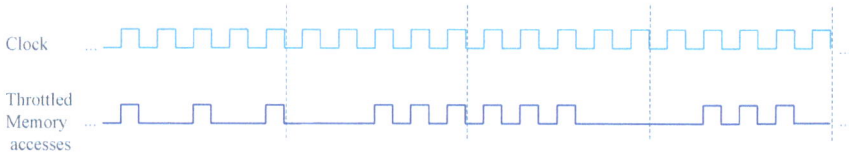

Fig. 1.18 Memory throttling that allows memory accesses within the time interval up to a quota. In this example, a quota of three memory accesses are allowed in each interval

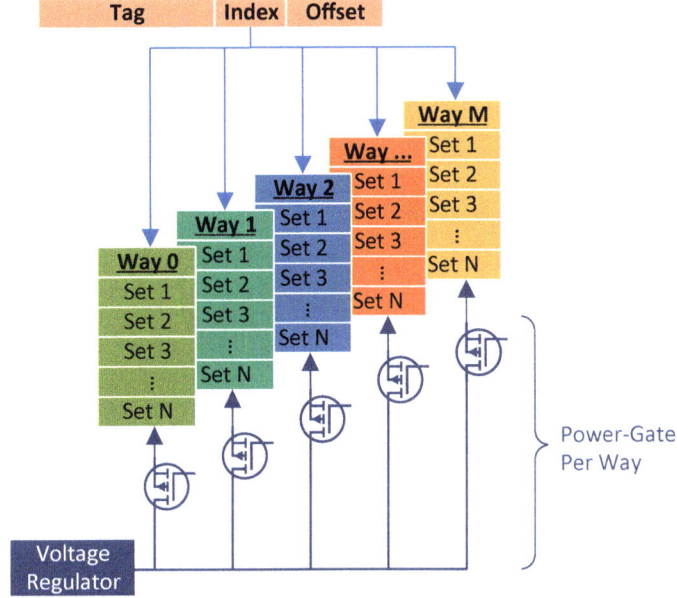

Fig. 1.19 Scheme for per-way cache resizing. Saving cache power when cache utilization is low

1.4 Power Management at Operating System

In this section, we give an overview of the system knobs that are used at modern processors in order to control the power and energy efficiency of the platform.

1.4.1 System Overview

Modern processors are built as SoCs, where each SoC integrates many components into a single chip. The integration can happen on the same silicon die or/and on package. Figure 1.20 shows a high-level scheme of modern processor's architecture. The main components are as follows:

- **Cores**: cores are execution units that are responsible for the execution of Instructions Set Architecture (ISA) code, the core fetches instruction from the instruction memory, executes them, and writes the results back to memory subsystem or to the registers. Recent processors include many cores (e.g., 2, 4, 10, and even more). Normally, the operating system is responsible for managing and scheduling tasks for the different cores.

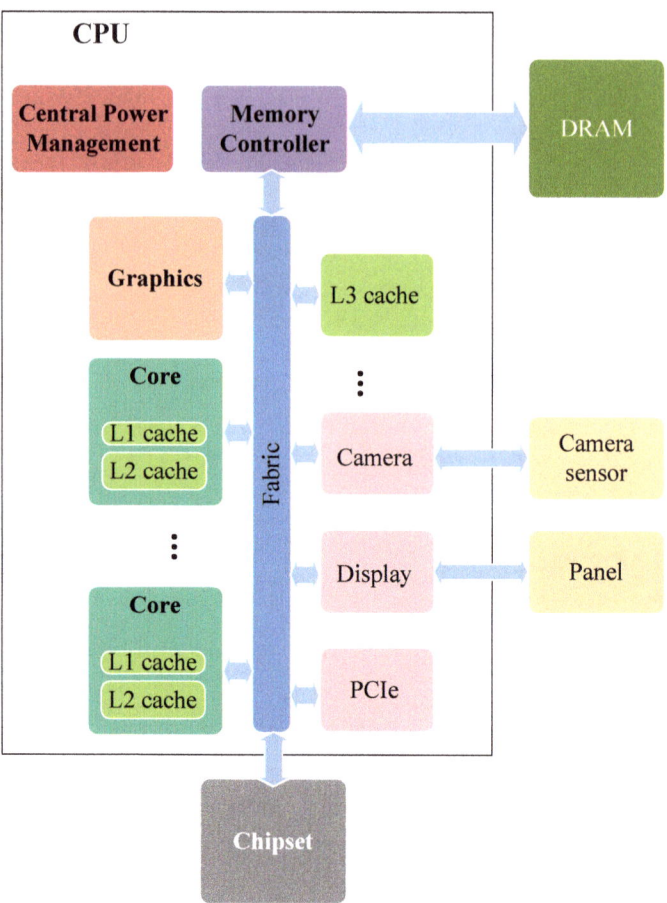

Fig. 1.20 High-level architecture of modern processors

- **Graphics**: many of today's processor integrate graphics engine into the SoC die, these engines share the memory subsystem with the other components and normally have high memory bandwidth demand. Graphics engines are normally managed by the graphics driver.
- **Caches**: caches are hardware component that store data closer to the processor, allowing future requests for that data to be served faster. A cache hit occurs when the requested data is found in a cache, while a cache miss occurs when it is not. Processor normally has three level of caches: First-Level Cache (FLC), also known as L1, Mid-Level Cache (MLC), also known as L2 and Last-Level Cache (LCC) or L3.
- **Memory controller**: The memory controller (MC) is a hardware component that manages the flow of data going to and from the main memory, known as Dynamic Random-Access Memory (DRAM). In modern processors, a memory

controller is integrated into the processor's die and usually called an integrated Memory Controller (iMC). Memory controllers implement the logic necessary to read and write to DRAM, and to "refresh" the DRAM.

- **Fabric**: Fabrics are the on-chip connection technologies that are vital to complex SoC implementation, where several cores, graphics, IPs, and memory are integrated together. Noncoherent fabrics are ideal for Direct Memory Access (DMA) usages, for example, display engine or camera processing unit accesses the DMA. Coherent fabrics are responsible for maintaining the shared data coherent between the connected components, for example, the coherent fabric between cores.
- **DRAM**: DRAM is a type of random-access memory, it is used mostly as a main memory as it has high memory density.
- **Intellectual Property** (IP): The present trend in SoC design is to make use of existing IPs as much as possible. IPs in the form of core, controllers, and accelerators are being reused in systems design in order to reduce Time To Market (TTM).
- **Central Power Management**: In modern processor, a central power management is implemented in order to handle the power, energy, and thermal management of the SoCs. All the global decisions on idle state, active state, and thermal events handling are taking place at this unit. Central PMU receives input from many sources, such as operating system, thermal sensor, current sensors, and performance counters. Moreover, the unit collects status data from local power management units that are distributed inside the SoC.

Modern processors expose interfaces to software and operating system for controlling processors features (e.g., frequency), in addition the processor gives observation to software developer on the performance, power, and micro-architectural events using performance-monitoring counters.

1.4.2 Advanced Configuration and Power Interface (ACPI)

Many Power Management algorithms and techniques, such as timeout-based as well as other predictive and stochastic policies, can be implemented in the hardware or firmware as a part of the SoC design. However, a hardware or firmware implementation highly complicates the modification and reconfiguration of the policies. Therefore, there are strong reasons to shift part of the implementation to the software level in order to give flexibility to operating system and Original Equipment Manufacturers (OEMs) to control and configure the power management at the platform. To address this problem Intel, Microsoft and Toshiba have published the first version of the Advanced Configuration and Power Interface (ACPI [42]) specification—an open standard that operating systems can use to discover and configure computer hardware components to perform power management. This standard defines a unified operating system-centric device configuration and power management interfaces. ACPI describes platform-independent interfaces for hardware discovery, configuration, power management, and monitoring.

ACPI is an attempt to unify and improve the existing power management and configuration standards for hardware devices. The standard brings power management into the operating system control and requires an ACPI-compatible operating system to take over the system and have the exclusive control of all aspects of the power management and device configuration responsibilities. The main goals of ACPI are to enable all computing systems to implement DPM capabilities, and simplify and accelerate the development of power-managed systems. It is important to note that ACPI does not put any constraints on particular power management policies, but provides the interface that can be used by software developers to leverage flexibility in adjustment of the system's power states.

ACPI defines a number of power states that can be applied in the system in runtime. The most important states are *G-states, S-states, C-states, P-states T-states, and D-states* as shown in Table 1.3.

1.4.2.1 Global States—G-States (Gx)

The ACPI specification [42] defines global system states (G0-G3) (Table 1.4):

1.4.2.2 Sleeping States—S-States (Sx)

Operating system can put the platform into one of the sleep states, the trigger for this process is configurable, it can be initiated after the platform being idle for long time or the user has pressed the power button. The OS puts the computer platform into one of the sleeping (Sx) states as defined by the operating system policy. No user-visible computation occurs in a sleeping state. OS can choose to implement or expose part of the states to the user. The main S-state is described in Table 1.5.

1.4.2.3 CPU States—C-States (Cx)

Under the working state of G0/S0, the ACPI defines various CPU power states. While each manufacturer can define the state and the behavior of the CPU at each state, here are some of the state recommended by ACPI (Table 1.6):

It is important to notice that these states are defined by the ACPI standard while the actual implementation by processors' manufacturers might be different. For example, Intel architecture defines various C-state levels [45, 46]: thread C-state (TCx), Core C-state (CCx), and Package level C-state (PCx). The operating system controls each core individually, while a PMU coordinates the cores and threads in order to enter the system to package C-state.

Table 1.3 ACPI processor power states

State	Symbol	Granularity	Description
G-states	Gx	Platform	Global power states of the platform, they determine the whole platform idle/off state and which reset type should be used when exiting the state into an active state
S-states	Sx	Package	Sleep states of the entire package, these states define the mode of sleep for the package, the states are initiated by the OS, the most popular state are S3 (suspend to memory) and S4 (suspend to disk or hibernate)
C-states	TCx/ CCx	Thread/core	C-states are idle power states of the threads/cores, respectively. These states are activated when the thread/core is idle, different levels of power and enter/exit latency exist for C-states while the trade-off is lower power with higher enter/exit latency, for example, CC1 has the highest power level with the shortest exit latency. On the other hand, CC6 has the lowest power level with the longest exit latency
Package C-states	PCx	Package	Package C-state is idle power state of the processor package, it is activated when all the compute domains (Cores and Graphics) are idle. Various levels of package C-states exist, ranging from clock gating at uncore level until nearly full shutdown of the package, similar to C-state, deep package C-states have lower power with higher exit latency and vice versa
P-states	Px	Package/ core (PCPS)	P-state changes the voltage and frequency of the cores (DVFS), traditionally the cores have one P-state domain while with PCPS feature that exists in some of the modern processors, it is possible to choose different P-state for different cores. Additional domain in the processors SoC has also the capability for P-state, such are the Last-Level Cache, the memory controller + DRAM and graphics, traditionally, the frequencies of these domains are not controlled directly by the OS, they are controlled by the PMU or driver
T-states	TTx	Core	These states are the core throttling, they are mainly used for aggressive reduction in core power (and performance) in cases of critical thermal events
D-states	Dx	Device	These states determine the power states of the devices, such as USB, SATA, PCIe, etc.)

C-States Example—Intel Architecture

Thread C-state (TC0-TC10)—When OS has no more tasks to schedule for a thread, the thread receives the idle task, this task executes MWAIT instruction with an argument that effectively denotes the thread C-state (TCx) that the thread will enter. The C-state level is selected according to OS algorithm based on the utilization history of thread/core.

Table 1.4 ACPI global states

G-state	Description
G0	Working state—This state also called S0 state
G1	Sleeping states—There are various level of sleep states (S1-S4)
G2	Soft-Off state—corresponding to S5 sleep state
G3	Mechanical Off state, normally only the Real-Time Clock (RTC) is working that is powered by onboard independent battery

Table 1.5 ACPI sleeping states

S-state	Description
S0	Working state
S1	Power-On and suspended. Normally the processor caches are flushed, and the cores stop executing instructions. The main power to the processor and memories (DRAMs) is maintained. Some of the devices can be turned off
S2	Processor power-off state. Caches will be flushed to main memory
S3	Standby state. OS state is suspended to main memory (RAM) The processor is turned off, main memory remains on at deep idle power state (e.g., memory self-refresh state)
S4	Hibernation state known also as suspend to disk state. The OS state is copied into the disk, the processor and memory (RAM) are turned off
S5	Soft-off state. All off except that the PSU or battery still supplies power, at a minimum level, to the power button to allow return to working state (S0). From this state, a full reboot is required. While no previous content is retained at memory of disk

Table 1.6 ACPI CPU states

S-state	Description
C0	Working state. CPU executes instruction. Under this state, the performance states (P-states: P_0-P_n) are defining the performance level at which the CPU cores run
C1	Known as Halt state. The CPU stop executing instructions and can return to running instruction almost instantly as the exit time to C0 is very short
C2	Known as Stop-Clock state. At this state, the processor turns off the clocks while the state is maintained
C3	known as CPU Sleep state. This is a state where the processor flushes the caches (and potentially turn them off). Additional power savings actions can be taken

At the thread level, minimal power actions are taken (e.g., clock gating of none-shared resources). The important power actions are taken at core and package C-states.

Core C-state (CC0-CC10)—Core C-state determines the idle power management action taken at core level. Under normal execution, a core is located at Core C-state C0 (CC0), while from CC1 to CC10, the core is idle and different power saving techniques are applied. The trade-off between the different states is power versus latency, thus the deeper the state is (higher CCx), the lower is the power consumed by the core while having higher latency to enter and exit the state. It is important to notice that some C-states apply the same power actions at core level, while these states determine to which package C-state the OS intends to enter at the end. For example, Core C6 (CC6) to C10 have the same power actions at core level (Core power gating) while the actions taken at package level are different. In addition, some of the new C-state does not exist at all processors families, while some states are no longer defined for new processors, for example the CC8–CC10 was first presented at Haswell processor [94], while CC2, CC4, and CC5 C-state are not defined at recent processors (Table 1.7).

Note: at Intel architecture, the Graphics (GPU) has its own idle state, known as **Render C-States** (RCx), **RC0** is the active state, and **RC6** is the idle state where the GPU is idle, clocks are turned off, and its voltage is lower or zero (via Power-gating or voltage regulator control). At the entry to this state, the internal state of the Graphics is saved in memory outside the Graphics and restored on wake-up.

Package C-state (PC0-PC10)—When the compute domains, Cores and Graphics(GPU), are idle, the processor has an opportunity for additional power savings at uncore[2] and platform levels, for example, flushing the LLC and power-gating it, power-gating the memory controller and DRAM IO, and at some state, the whole processor can be turned off while its state is preserved on always-on power domain.

Package C-states are targeted at idle or close to idle conditions. The exact actions taken at each package C-state change from processor to another. However, the high-level concept remains the same. At client processors, the following cases denotes different power states:

- Compute domains are idle, path to memory open
- Path to memory closed, uncore is clock gated, display is ON
- Path to memory closed, uncore power-gated, display is ON
- Path to memory closed, uncore power-gated, display is OFF

The high-level definition of the Package C-state at Intel architecture is as follows (Table 1.8).

1.4.2.4 Performance States—P-States (Px)

When the processor's core is active (operates at CC0), it can be in one of several power performance states (P-states). These states are implementation-dependent. The ACPI defines that the P0 is always the highest performance state; with P1 to Pn being successively lower performance states.

---

[2]Uncore is a term used by Intel to describe the functions of a microprocessor that are not in the core or Graphics, but which must be closely connected to the core to achieve high performance, for example the memory controller.

Table 1.7 Example of core C-state at Intel architecture

Core C-state	Wake-up latency[a]	Description
CC0	N/A (running state)	The running or active state. At this state, one or more threads are executing instruction. Autonomous clock gating is may be applied for unused logic blocks
CC1	<1us	In CC1, the core main clocks are gated. The dynamic power is driven close to zero. Core caches and TLBs are maintained coherent
CC1E	<1us + DVFS transition (if required)	Enhanced CC1—hint to drop voltage: CC1E is effectively the same as C1, except that it provides a hint to the global voltage/frequency control that frequency/voltage can be reduced to save additional power. The additional power savings over CC1 are coming from static power as the voltage is reduced (along with DVFS change)
CC3	50–80 us	Clock sources (PLLs) off, cache flush and going to retention voltage. The core state is maintained on place, the voltage is dropped to retention voltage in order to preserve the core state inside the registers. L1, L2, and TLB are flushed
CC6	50–100 us	Power-gating state: The core is power-gated using on-die Power Gates. L1 and L2 caches are flushed. Core TLBs are flushed. At the entry to this C-state, the internal state of the core is saved in memory outside the core and restored on wake-up
CC7-C10	50–100 us	For CC7–CC10, the power savings actions for the core are similar to CC6, while additional power savings actions will be taken outside the core at package C-states

[a]These numbers describe the exit latency at Core level, while system-level exit latency depends on the package C-state that was entered at the end. The exit latency is implementation-dependent, and is normally faster at recent processors, for example, processors with IVR will have faster exit latency for some of the states

- P0 has the maximum power and frequency state
- P1 has less power and frequency than P0, voltage and frequency scaled
- P2 has less power and frequency than P1, voltage and frequency scaled
- …
- Pn has less power and frequency than Pn-1, voltage and frequency scaled

In recent Intel processor, P-states is controlled by the SpeedStep technology and algorithms that attempt to raise the energy efficiency of the processor by wisely adjusting the processor P-state while taking into account different parameters, e.g., performance scalability, where performance scalability is performance and frequency correlation; a ratio of 1 means doubling frequency results in doubling performance.

P-states were invented in order to dynamically adjust the processor operating voltage and frequency to match the needs of the user at a given point in time.

Table 1.8 Example of package C-state at Intel architecture

Package C-state	Core/render C-states	Path to memory	Display	Description
PC0	At least one in CC0/RC0	Available	Available	The running or active state. At this state, at least one Core or Graphics is working
PC2	All cores in CC3+ , GPU at RC6	Available	Available	All Cores entered to CC3 or deeper C-states, Graphics at RC6. Path to memory is open as some IOs requesting access to memory (e.g., Camera accelerator)
PC3	All cores in CC3+ , GPU at RC6	Not available	Available	All Cores entered to CC3 or deeper C-states, Graphics at RC6. Path to memory is closed and memory enters Self-Refresh state
PC6	All cores in CC6+ , GPU at RC6	Not available	Available	All Cores entered to CC6 or deeper C-states, Graphics at RC6. Path to memory is closed and memory enters Self-Refresh state
PC7	All cores in CC7+ , GPU at RC6	Not available	Available	All Cores entered to CC7 or deeper C-states, Graphics at RC6. Path to memory is closed and memory enters Self-Refresh state. Additional power gating at uncore is done
PC8	All cores in CC8+ , GPU at RC6	Not available	Available	All Cores entered to CC8 or deeper C-states, Graphics at RC6. Path to memory is closed and memory enters Self-Refresh state. Additional power gating at uncore is done (e.g., IOs)
PC9	All cores in CC9+ , GPU at RC6	Not available	Not available	All Cores entered to CC9 or deeper states, Graphics at RC6. Path to memory is closed and memory enters Self-Refresh state. Uncore is power-gated including Display. Additional power savings at platform level are carried out
P10	All cores in CC10, GPU at RC6	Not available	Not available	All Cores entered to CC10 C-states, Graphics at RC6. Path to memory is closed and memory enters Self-Refresh state. Uncore is power-gated including display. Additional power savings at platform level are carried out. This state is mostly used as the connected standby state

Running at lower frequencies results in lower performance and longer latency to complete the same amount of work. However, it may be possible to complete a required amount of work with lower energy. A good example is a processor that is running light workload such as video playback, the processor's utilization is relatively low ($\sim 10\%$), in this workload, the processor can run at the minimum frequency (Pn). As the system load begins to increase (e.g., user opened new application), the frequency can be increased to meet the higher level of demand while a continued quality of service is maintained. The operating system has traditionally been responsible for selecting which frequency the system should operate at.

While the ACPI defines the P-state as a standard performance states, each processor's vendor has his own implementation and algorithm on top of P-states, for example Intel defined the Turbo and SpeedShif to manage the P-state transitions for modern processors. In Intel's processors, DVFS is controlled by choosing performance state (P-state), this is done by writing to per-thread Model-Specific Register (MSR) accessed by the operating system kernel. The PMU considers the requests from all running threads and sets the frequency to the maximum frequency requested by all of the threads. The requested P-states are values between P0 and Pn, where P0 is the maximum frequency with Turbo enabled, P1 is the maximum guaranteed frequency, and Pn is the lowest available frequency. For example, at one of the systems, the P0 is 2.8 GHz, P1 is 1.9 GHz, and Pn is 0.8 GHz. The DVFS request and the Turbo control are done by writing to the IA32_PERF_CTL MSR.

Turbo Boost

Turbo Boost—also known as Intel Turbo Boost Technology 2.0, or Speed Step, introduced with Intel's 2nd Generation Core™ processor—opportunistically boosts the frequencies of the cores in multi-core Intel Processors. The processor hardware controls the Turbo Boost activation and the level of boosting depends on the number of active cores, estimated power consumption, and the temperature of the package. This "thermal boosting" allows the processor to temporarily exceed the Thermal Design Point (TDP) using the thermal capacitance of the package. The activation is done when the OS request P0, thus all the frequency levels at the range between P0 and P1 are autonomously controlled by the processor when P0 is requested by the OS.

Figure 1.21 illustrates the Turbo behaviors over time. In the first phase (at C-state), the processor gains thermal budget while sleeping (e.g., deep Package C-state). In the second phase (C0/P0), the processor moves to Turbo (P0 in the P-state terminology [55]) following a P0 request by the software. In this stage, the processor starts to consume its thermal budget and at a later stage as the processor heats up the processor's hardware starts to reduce the frequency, the processor temperature gets close to the maximum allowed temperature. Once the entire thermal budget is consumed, the processor's frequency normally stabilizes at the frequency corresponding to TDP frequency (P1 guaranteed frequency). The processor is not allowed to go back to Turbo until a new thermal budget is accumulated.

Turbo Boost might be energy inefficient in some cases. To cope with this problem, Intel's modern processors feature the ability for software to control the energy efficiency of the processor by configuring the IA32_ENERGY_PERF_BIAS register [55].

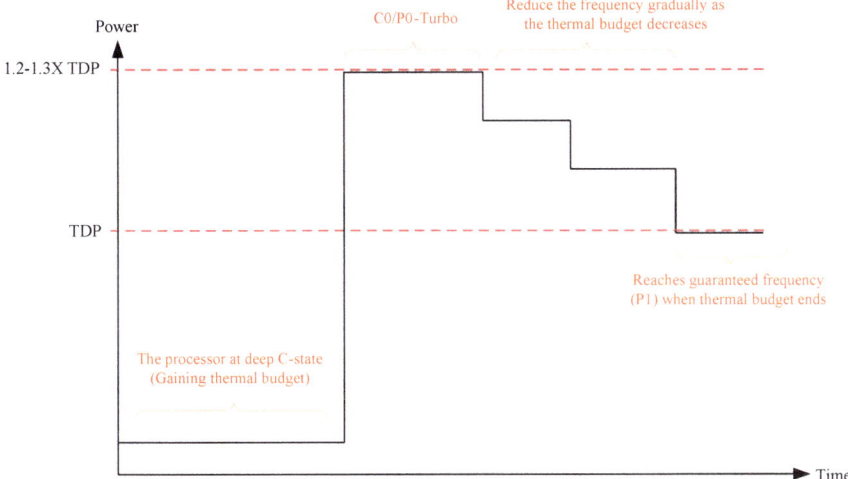

Fig. 1.21 Turbo boost behavior overtime, thermal budget gained at low-power intervals is used to temporarily run above TDP frequency

Speed Shift

Speed Shift [55]—introduced in the 6th Generation Core processors, codename Skylake—redefines the processor-OS performance states interface. While at Turbo Boost, the processor is responsible for controlling the frequencies between P0 and P1, with Speed Shift, the OS predefines a frequency range for the processor to manage the instantaneous frequency at runtime.

Compared to Turbo Boost, Intel's new Speed Shift terminology changes the responsibilities of controlling the P-state by having the operating system moving some or all control of the P-States handing to the processor. This feature has various advantages, as it is much faster for the processor (Hardware) to control the ramp up and down of the frequency, compared to OS (Software). In addition, the processor has much finer control and visibility (e.g., fast access to performance counters, thermal sensors, etc.) over its states, allowing it to choose the most optimum performance level for a given timeframe, and therefore using less energy as a result. With Speed Shift, frequency transition time is reduced to around 1 ms with Speed Shift controlled processor compared to 20–30 ms on OS controlled system. Moreover, the transition from Pn P-state to maximum performance state (P0) can be done in around 30–40 ms, compared to around 90–100 ms with the Turbo Boost implementation.

As seen in Fig. 1.22, traditionally, the operating system controls the frequencies between P1 and Pn according to the ACPI tables, the OS is responsible for explicit selection of the P-state, when the OS selects P0, it offloads the P-states control to the PMU at hardware (or firmware), the PMU at hardware selects the running frequency between P1 and P0 single core (P0-1 core). At this definition, the

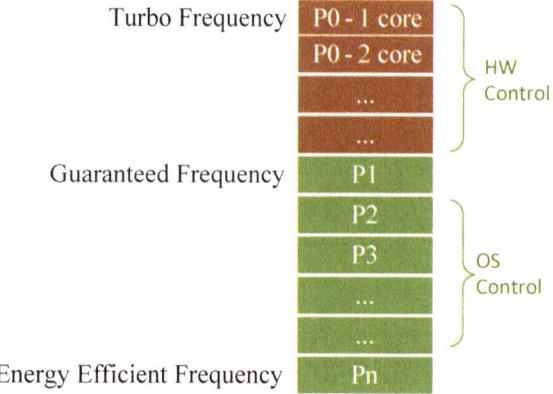

Fig. 1.22 P-state definition
before the speed shift

energy-efficient frequency was defined as the Pn frequency, which was the lowest frequency also, lower frequency was applied at critical conditions (such as thermal event).

Moving to Speed Shift, the definition of the P-state handling was changed as shown in Fig. 1.23. The highest frequency is still defined as the P0 1-core, while the guaranteed frequency is configurable. Pn at the legacy definition was marked as the energy-efficient frequency, this was correct for the compute domain only and not for the whole platform, a new point called the most efficient frequency (Pe) was defined with Speed Shift, this reflects the most energy-efficient frequency at platform level. Pe is calculated at runtime and takes into account many parameters such as temperature, graphics frequency, workload scalability, etc. At Speed Shift, the OS has the option to control the P-state at any point while at most cases, it gives the control to the hardware PMU to do autonomous frequency control for the whole frequency range. In addition, the OS has the option to set lower and upper limit of the allowed frequency where the hardware algorithm will work, and a new control

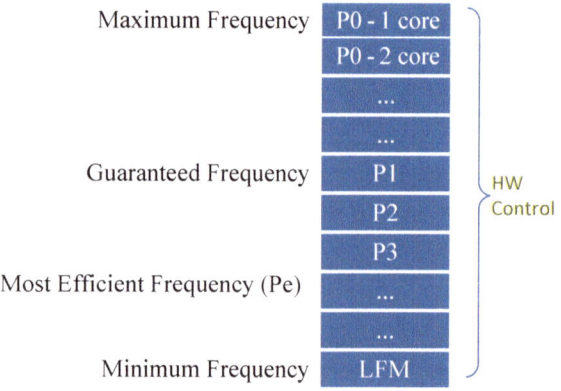

Fig. 1.23 P-state new
definition with speed shift

called Energy Performance Preference (EPP) was added where the OS can give directives to the hardware autonomous algorithm to set the energy versus performance preferences.

Per Core P-State (PCPS)

Traditionally, multi-core processors, such as Intel client processor, have one P-state domain for all core, this means that all active cores are running at the same frequency and voltage, while the P-state point is chosen as the maximum P-state among all requests from all active cores.

Recently, server processors from Intel provide the ability to independently change the frequency and voltage of the individual cores in the processor, as known as, PCPS. In this mode of operation, the target frequency of a given core is the maximum of the requested frequency for the threads on the core.

Most of client workloads tend to have balanced utilization of the cores in average, while normally server has much more cores and workloads (such as web servers) that service small, discrete tasks. As the tasks come into the system, they are forked out to the various threads that service them. In this type of workload, different hardware cores tend to observe imbalances in utilization, at these workloads, it is normally more energy efficient to reduce the frequency and voltage (lower P-state) on the less utilized cores. Additional example where PCPS can be used is with virtualization at servers. Where one user may desire high utilization of their virtualized resources, whereas another may be running at very low utilization. PCPS allows the active user to ramp their voltage and frequency up while having minimal impact of on the other user.

The introduction of Integrated Voltage Regulator (IVR) in recent modern processors, such as FIVR at Intel processors, has enabled the PCPS as it easily offers individual voltage domain for each core. LDOs can also be used to provide variable voltages across cores in a processor with a single input voltage, this technique is being used at AMD recent processors.

1.4.2.5   Device States—D-States (Dx)

The ACPI defines a state for devices, such as PCIe, SATA, USB, etc. The device states *D0–D3* are device dependent (Table 1.9):

Traditionally, according to ACPI, devices can be in active power state (D0) or in low-power state (D1/D2/D3). Most platforms implement D0 and D3, while in order for the device to enter to these states, the OS needs to request or to be aware of the device state, new device idle states were defined that are transparent to the OS and can be activated by the hardware or the device driver, there states are called D0ix states (i.e., idle state under the D0 state), where from OS perspective, the device is at D0 while from hardware point of view, the device is at idle state (D0ix) (Table 1.10):

Table 1.9 Devices states description

D-state	Description
D0	This is the active state for the device
D1, D2	low-power state of the device. These states are intermediate power states and their definition varies by device and vendors
D3	D3 if idle state that is further divided into D3 *Hot* and *Cold*: •*D3-Hot*: The device can assert power management requests to transition to higher power states, the device in this state will be idle while keeping wake-up logic alive that responses to requests from the bus or IOs •*D3-Cold*: This state is known also as the Off state, the device powered off and unresponsive to requests from bus or IOs

Table 1.10 Idle states under D0 (D0ix)

D-state	Description
D0i1	Clock-gated state on device and IO
D0i2	Keep internal state on retention voltage
D0i3	Most of the device components are power gated, keep wake-up logic to monitor the wake requests. Usually this state is controlled by the driver or the PMU of the device subsystem with no OS control of the subsystem. Normally, the device driver or the PMU coordinates and manages the subsystem state and the state saves/restores if needed for power transitions

1.4.2.6   Throttle States—T-States (Tx)

Throttling states are defined per core, these states perform duty cycling of the cores at a fixed interval, the states reduce the power and also the performance of the core, they are less efficient than DVFS (P-state). T-states are generally used for aggressive throttling when needed for thermal, electrical, or power reasons. They traditionally have not been used for power efficiency.

1.4.3 Performance Counters

Performance Monitor Counters (PMC) exist on most modern, high-performance processors; such as Intel recent processors; PMC is special, on-chip hardware that monitors processor's performance. Data collected by this hardware provides performance information on applications, the operating system, and the processor. These data can guide performance improvement efforts by providing information that helps programmers tune the algorithms used by the applications and operating system, and the code sequences that implement those algorithms.

Intel processors provide performance-monitoring counters for monitoring internal hardware operations (e.g., unhalted core cycles, instructions retied, cache misses) as shown in Fig. 1.24. Configuring an architectural performance-monitoring event involves programming performance event select Model-Specific Register (MSR). There are a finite number of performance event select MSRs (IA32_PERFEVTSELx MSRs). The result of a performance-monitoring event is reported in a performance-monitoring counter (IA32_PMCx MSR). Performance-monitoring counters are paired with performance-monitoring select registers.

The Read Performance Monitoring Counter (RDPMC) instruction allows application programs to read the processor's performance-monitoring counters.

1.4.3.1 Performance-Monitoring Hardware

Performance-monitoring hardware typically has two components: performance event detectors and event counters. By properly configuring the event detectors and counters, users can obtain counts of a variety of performance events under various conditions. Users can configure performance event detectors to detect any one of several performance events (for example, cache misses, vector instruction executed). Often, event detectors have an event mask field that allows further qualification of the event. For example, at Intel processors [80]; event to count load accesses to LLC has an event mask that lets event detectors monitor only accesses to cache lines in a specific state—modified, shared, exclusive, or invalid.

In addition to counting events detected by the performance event detectors, users can configure performance event counters to count only under certain edge and threshold conditions. The edge detection feature is most often used for performance events that detect the presence or absence of certain conditions every cycle. For these events, an event count of one represents a condition's presence and zero indicates its absence. For example, a pipeline stall event indicates the presence or absence of a pipeline stall on each cycle. The event counter's second major feature is threshold support. This capability lets the event counter compare the value it reports each cycle to a threshold value. If the reported value exceeds the threshold, the counter increments by one. The threshold feature is only useful for performance events that report values greater than one each cycle. For example, superscalar processors can complete more than one instruction per cycle. Selecting instructions completed as the performance event and setting the counter threshold to two would increment the counter by one whenever three or more instructions complete in one cycle. This provides a count of how many times three or more instructions completed per cycle.

1.4.3.2 Performance Event Monitoring

Performance events can be grouped into five categories:

- Program characterization
- Memory accesses
- Pipeline stalls

Fig. 1.24 Performance
counter at Intel architecture,
there are performance counter
per core, and other
performance counter at
processor level (e.g., uncore)

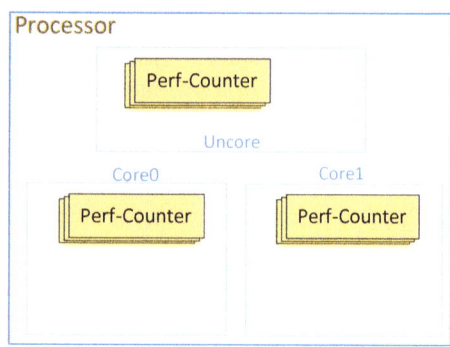

- Branch prediction
- Resource utilization

Program characterization events help define the attributes of a program (and/or the operating system) that are largely independent of the processor's implementation. The most common examples of these events are the number and type of instructions (for example, loads, stores, floating point, vector, branches, and so on) completed by the program.

Memory access events often comprise the largest event category and aid performance analysis of the processor's memory hierarchy. For example, memory events can count references and misses to various caches and transactions on the processor memory bus.

Pipeline stall event information helps users analyze how good program's instructions flow through the pipeline. Processors with deep pipelines rely heavily on branch prediction hardware to keep the pipeline filled with useful instructions.

Branch prediction events let users analyze the performance of branch prediction hardware (for example, by providing counts of mispredicted branches).

Resource utilization events let users monitor how often a processor uses certain resources (for example, the number of cycles spent using a floating-point divider). Table 1.11 shows some performance counters events and its description from Intel Architecture.

1.4.4 Energy Efficiency Versus Performance Control at Modern Processors

As the energy consumption and battery life importance is growing, performance and power metrics alone are not enough, several new metrics are presented that quantify the efficiency of the processor. At this subsection, we list some of them. In addition, we demonstrate several knobs that exist at modern processors for controlling the energy versus performance preferences.

Table 1.11 Examples of performance counters in Intel Skylake processors

Performance counters	Description
UOPS_DISPATCHED_PORT.PORT_1	Arithmetic operations dispatched
UOPS_DISPATCHED_PORT.PORT_2	Memory operations dispatched
UOPS_DISPATCHED_PORT.PORT_6	Branch operations dispatched
IDQ.DSB_UOPS	Micro-operations (Uops) delivered by the decoded instruction cache (also known as DSB)
IDQ.MITE_UOPS	Uops delivered to the traditional instruction fetch and decode pipeline
IDQ.MS_UOPS	Uops delivered microcode sequencer (responsible to deliver complicated instruction with long sequences like CPUID)
LSD.UOPS	Uops delivered by the Loop-Stream Detector
L2_RQSTS.MISS	Memory access that miss in the 2nd level cache (L2)
FP_ARITH_INST_RETIRED_256BIT_PACKED_DOUBLE	Floating-point instructions with 256 bits vector double precision
FP_ARITH_INST_RETIRED_256BIT_PACKED_SINGLE	Floating-point instructions with 256 bits vector single precision

1.4.4.1 Energy Per Instruction (EPI)

One of energy efficiency metrics is the Energy Per Instruction (EPI); EPI is sometimes used as a method of comparing energy optimizations, particularly those that focus on general micro-architectural characteristics, rather than on the runtime of a particular application.

1.4.4.2 Energy-Delay Product (EDP)

With the trend of going low in power consumption while maintaining performance, the metric of EDP [81] was presented, this metric takes into account the energy consumption of a workload and the time it talks to complete it. The EDP gives equal "weight" for the performance and power.

1.4.4.3 Energy-Delay-Squared Product (ED²P)

Energy-Delay-Squared Product is a metric similar to EDP but gives more "weight" for the performance over the power. This metric, and additional metrics like ED^3P, gives the performance more importance over energy.

1.4.4.4 Energy Efficiency Control with Turbo Boost

Intel's modern processors feature the ability for software to control the energy efficiency of the processor by configuring the IA32_ENERGY_PERF_BIAS register [82]. The processor can be configured to favor highest performance, maximum energy savings or a value in-between.

1.4.4.5 Energy Efficiency Control with Speed Shift

With Speed Shift, the operating system can determine the energy efficiency parameters versus the required performance by writing to a MSR. The operating system controls Speed Shift operation for each processor via writing of control hints/constraints to the IA32_HWP_REQUEST MSR at the Energy_Performance_Preference (EEP) field. The operating system may write a range of values from 0 (performance preference) to 0FFH (energy efficiency preference) to influence the rate of performance increase/decrease and the result of the hardware's energy efficiency and performance optimizations.

1.5   Power Estimation Techniques

Power estimation techniques can be divided into two categories: runtime power estimation and offline power estimation.

1.5.1   Runtime Power Estimation

Several proposals either aim to estimate runtime power or might seem suitable for runtime power estimation.

1.5.1.1 Current Sensors

The first runtime power monitoring category targets power estimations using a current sensor as one of the inputs. Isci and Martonosi [56] implement runtime power monitoring for a single-core Intel® Pentium™ 4 (Willamette) using a combination of performance counters and an inline external ammeter. Performance-counter results and current measurements are fed to a separate monitoring machine which computes power in software on intervals of about 400 ms, the model estimates power for 22 microprocessor structures using 24 statistics plus the meter. While this model is quite useful, the external hardware requirement for model construction makes it unwieldy for runtime power monitoring in the field, particularly in a multi-core microprocessor where each processor core might require an on-package current monitor.

More recently Intel started to use the current sensor for power and energy estimation at its processors. Starting from Haswell processor, the external voltage regulator reports its current consumption to the PMU periodically and the PMU calculates the power and energy consumed since last reporting. The power and energy data that the PMU calculates is used by the PMU power management algorithms. In addition, the energy reading is accumulated into the Running Average Power Limit (RAPL) Model-Specific Register (MSR), which can be read by the user with privileged kernel mode. These energy counters give visibility to the energy consumption of the package, cores, graphics (in client processor) and memory domains (in server processor).

RAPL data can be configured and examined by reading MSRs. On the Intel architecture, today this is only possible in privileged kernel mode. The available RAPL energy counters at the Skylake processor are listed at Table 1.12.

1.5.1.2 Low-Level Hardware Events

Other power estimation proposals rely only on performance metrics as input statistics. One category uses a large number of wire-specific or structure-specific statistics relative to the number of structures for which power is estimated. Examples include the Cai-Lim mode [57], which uses structure activity and power densities to compute power for 17 structures and PowerTimer, which relies on many switching factors as input statistics. Another example is Wattch [58], which is

Table 1.12 List of available energy telemetries at Intel processors

Domain	Description
PKG_ENERGY	Energy of the whole processor package
PP0_ENERGY	Energy of power Plane 0 domain (Processor IA cores domain and LLC)
PP1_ENERGY	Energy of power Plane 1 domain (specific to client processors where it is the integrated processor graphics)
DRAM_ENERGY	Energy of the DRAM domain (specific to server processors)
PLATFORM_ENERGY	Energy of the entire platform including IA core, Integrated graphics and the system agent and other platform components. (new in Skylake)

similar to the Cai-Lim model but derives energy costs from wire delay and circuit models instead of power density. Wattch uses a few tens of input statistics to estimate power for about a dozen structures. Wu et al. [59] estimate power for 15 Pentium 4 structures using up to 22 input statistics, and Peddersen et al. [60] estimate power for 5 structures in a small embedded-type core. While these estimation techniques are conceptually suitable for runtime power estimation, the number of input statistics relative to the number of structures for which they compute power makes them impractical for runtime estimation for many structures. Sandy Bridge [66] modern processor implements runtime energy estimation using architectural events. The PMU collects a set of architectural events from each Intel architecture core, the processor graphics, and the I/O, and combines them with energy weights to predict the package's active power consumption. Leakage information is coded into the die and is scaled with operating conditions such as voltage and temperature to provide the package's total power consumption. The system uses architectural power predictor output, which is also exposed externally to software to decide the amount of turbo upside available for the current workload.

1.5.1.3 Performance Counters

Another category of power estimators relies on a handful of global or generic input statistics for core-level estimates, and thus would be more suitable for runtime power estimation. One example is Joseph and Martonosi's work [62], which uses performance counters to estimate power in the Intel® Pentium Pro™ over 10 ms intervals. The authors note that there is limited clock gating in this microprocessor and thus little variation between minimum and maximum power (about 25%). Estimating runtime power on a complex wide-issue processor with aggressive clock gating is substantially more difficult, as noted by Isci and Martonosi [61]. Due to limited clock gating, Joseph and Martonosi [62] were able to assume constant power for many structures and primarily focus on memory and ALU operation power. Another work, Bellosa [65], estimates core power at runtime in software, using Intel® Pentium™ 4 performance counters. Sharkey [63] also estimates core-level power. Powell et al. [64] proposed a technique, called CAMP that uses global input statistics to estimate core or die power. CAMP estimates per-structure power for over 180 structures inside Intel-like microprocessor.

1.5.2 Offline Power Estimation

There are a number of analytical power models that target offline power estimation using inputs that would be unsuitable for runtime estimation in hardware.

1.5.2.1 Profile-Based, Sampling, or Probabilistic

One category uses profiling, sampling, and statistical or probabilistic analysis to estimate power. The typical approach is to perform detailed simulations of a few design points, and use data from these to fit inference models that relate power to architectural statistics and then predict the power at different design points [68–70]. Lee and Brooks [68] used trace-driven simulation of 4000 samples to fit linear regression models that relate several micro-architectural design parameters to performance and full CPU power. Similar analysis can be done at the Register Transfer Level (RTL) level, such as Macii et al. [70], which uses statistical sampling for RTL estimation and Katkoori et al. [72], which generates behavioral profiles from RTL simulations.

1.5.2.2 Wire or Structure-Specific Statistics

Offline estimation at the circuit level or RTL can also use wire-specific or structure-specific inputs (e.g., temperature estimates or transistor count) for even finer granularity estimations. For example, Srinivasan et al. [72] present an empirical approach to determine the optimal pipeline depth considering both power and performance constraints. Zyuban and Strenski [73] present a mathematical approach based on hardware intensity, which relates delay and average energy consumption. Lee et al. [74] analyze low-level power for embedded DSP software. Landman et al. [75] survey power estimation techniques ranging from counting gate equivalents to circuit-level estimation. PowerTimer [77] and Wattch [76] can also be considered in this space when specific values of input bits (e.g., the addends in an addition operation) are considered as one of the inputs.

1.5.2.3 Analytical Estimation at Design

Analytical performance models are related to offline power estimation techniques. These models might be used to provide estimates for the input parameters to the power model before an architectural simulator is available. Many analytical performance models determine overall performance for an ideal processor (no misses, infinite hardware resources) and then derive performance limits imposed by adding constraints, such as dependencies and hardware limitations. The models are typically parameterized with data obtained from trace-driven simulations. Noonburg and Shen [79] develop such a model based on probability matrices, while Karkhanis and Smith [78] present a more concise model that calculates performance using two components—a constant, ideal component, and a performance loss component.

References

1. NASA Goddard Institute for Space Studies (n.d.). GISS Surface Temperature Analysis. Accessed November 30, 2010.
2. NRE (2010) National renewable energy laboratory. URL www.nrel.gov.
3. de Vries RP (2010) Green chips: A new era for the semiconductor industry. In: Proc. IEEE 2010 Custom Integrated Circuits Conf.
4. EER (2010) The office of energy efficiency and renewable energy. URL www.eere.energy.gov.
5. TOP 500 SUPERCOMPUTER SITES, http://www.top500.org/list/2013/06 (accessed December 12, 2013).
6. THE GREEN500 SITES, http://www.green500.org (accessed December 12, 2013).
7. Barroso, L.A., Clidaras, J. and Hölzle, U., 2013. The datacenter as a computer: An introduction to the design of warehouse-scale machines. Synthesis lectures on computer architecture, 8(3), pp. 1–154.
8. M. Poess and R. O. Nambiar, "Energy cost, the key challenge of today's data centers: A power consumption analysis of TPC-C results," Proc. VLDB Endowment, vol. 1, no. 2, pp. 1229–1240, Aug. 2008.
9. R. Buyya, C. Vecchiola, and S. Selvi, Mastering Cloud Computing: Foundations and Applications Programming. Amsterdam, The Netherlands: Elsevier, 2013.
10. Y. Gao, H. Guan, Z. Qi, B. Wang, and L. Liu, "Quality of service aware power management for virtualized data centers," J. Syst. Architect., vol. 59, no. 4/5, pp. 245–259, Apr./May 2013.
11. S. Rivoire, M. Shah, P. Ranganathan, C. Kozyrakis, and J. Meza, "Models and metrics to enable energy efficiency optimizations," Computer, vol. 40, no. 12, pp. 39–48, Dec. 2007.
12. K. Bilal, S. Malik, S. Khan, and A. Zomaya, "Trends and challenges in cloud datacenters," IEEE Cloud Comput., vol. 1, no. 1, pp. 10–20, May 2014.
13. B. Whitehead, D. Andrews, A. Shah, and G. Maidment, "Assessing the environmental impact of data centres—Part 1: Background, energy use and metrics," Building Environ., vol. 82, pp. 151–159, Dec. 2014.
14. V. Mathew, R. K. Sitaraman, and P. J. Shenoy, "Energy-aware load balancing in content delivery networks," CoRR, vol. abs/1109.5641, 2011.
15. P. Corcoran and A. Andrae, "Emerging trends in electricity consumption for consumer ICT," Nat. Univ. Ireland, Galway, Ireland, Tech. Rep., 2013.
16. Koomey, J., 2011. Growth in data center electricity use 2005 to 2010. A report by Analytical Press, completed at the request of The New York Times, 9.
17. Van Heddeghem, W., Lambert, S., Lannoo, B., Colle, D., Pickavet, M. and Demeester, P., 2014. Trends in worldwide ICT electricity consumption from 2007 to 2012. Computer Communications, 50, pp. 64–76.
18. "Energy efficiency policy options for Australian and New Zealand data centers," The Equipment Energy Efficiency (E3) Program, 2014.
19. Info-Tech, "Top 10 energy-saving tips for a greener data center," Info-Tech Research Group, London, ON, Canada, Apr. 2010.
20. S. Yeo, M. M. Hossain, J.-C. Huang, and H.-H. S. Lee, "ATAC: Ambient temperature-aware capping for power efficient datacenters," in Proc. ACM SOCC, 2014, pp. 17:1–17:14.
21. S. Devadas and S. Malik, A survey of optimization techniques targeting low power VLSI circuits,‖ in Proceedings of the 32nd ACM/IEEE Conference on Design Automation, 1995, pp. 242–247.
22. V. Tiwari, P. Ashar, and S. Malik, "Technology mapping for low power" in Proceedings of the 30th Conference on Design Automation, 1993, pp. 74–79.
23. Intel math kernel library (MKL). http://software.intel.com/en-us/articles/intel-mkl/.
24. J. Cebrian et al., Optimized hardware for suboptimal software: The case for SIMD-aware benchmarks, in ISPASS, 2014.

25. L. Benini, A. Bogliolo, and G. D. Micheli, "A survey of design techniques for system-level dynamic power management" IEEE Transactions on Very Large Scale Integration (VLSI) Systems, vol. 8, no. 3, pp. 299–316, 2000.
26. S. Albers, "Energy-efficient algorithms" Communications of the ACM, vol. 53, no. 5, pp. 86–96, 2010.
27. V. Venkatachalam and M. Franz, Power reduction techniques for microprocessor systems,‖ ACM Computing Surveys (CSUR), vol. 37, no. 3, pp. 195–237, 2005.
28. G. Buttazzo, Scalable applications for energy-aware processors,‖ in Embedded Software, 2002, pp. 153–165.
29. Efraim, R., Ginosar, R., Weiser, C. and Mendelson, A., 2014. Energy aware race to halt: A down to EARtH approach for platform energy management. IEEE Computer Architecture Letters, 13(1), pp. 25–28.
30. M. Weiser, B. Welch, A. Demers, and S. Shenker, Scheduling for reduced CPU energy,‖ Mobile Computing, pp. 449–471, 1996.
31. K. Govil, E. Chan, and H. Wasserman, Comparing algorithm for dynamic speed-setting of a low-power CPU,‖ in Proceedings of the 1st Annual International Conference on Mobile Computing and Networking (MobiCom 2005), Berkeley, California, USA, 1995, p. 25.
32. A. Wierman, L. L. Andrew, and A. Tang, Power-aware speed scaling in processor sharing systems,‖ in Proceedings of the 28th Conference on Computer Communications (INFOCOM 2009), Rio, Brazil, 2009.
33. L. L. Andrew, M. Lin, and A. Wierman, Optimality, fairness, and robustness in speed scaling designs,‖ in Proceedings of ACM International Conference on Measurement and Modeling of International Computer Systems (SIGMETRICS 2010), New York, USA, 2010.
34. A. Weissel and F. Bellosa, Process cruise control: event-driven clock scaling for dynamic power management,‖ in Proceedings of the 2002 International Conference on Compilers, Architecture, and Synthesis for Embedded Systems, Grenoble, France, 2002, p. 246.
35. K. Flautner, S. Reinhardt, and T. Mudge, Automatic performance setting for dynamic voltage scaling,‖ Wireless networks, vol. 8, no. 5, pp. 507–520, 2002.
36. S. Lee and T. Sakurai, Run-time voltage hopping for low-power real-time systems,‖ in Proceedings of the 37th Annual Design Automation Conference, Los Angeles, CA, USA, 2000, pp. 806–809.
37. J. R. Lorch and A. J. Smith, Improving dynamic voltage scaling algorithms with PACE,‖ ACM SIGMETRICS Performance Evaluation Review, vol. 29, no. 1, p. 61, 2001.
38. C.-H. R. Wu, "U.S. patent 7352641: Dynamic memory throttling for power and thermal limitations." Sun Microsystems, Inc., issued 2008.
39. I. Hur and C. Lin, "A comprehensive approach to DRAM power management," in Proceedings of the 14th International Symposium on High Performance Computer Architecture (HPCA'08), August 2008.
40. K.T. Sundararajan, T.M. Jones, and N. Tophamet, "Smart Cache: a self-adaptive cache architecture for energy efficiency," Int. Conference on Embedded Computer Systems: Architectures, Modeling, and Simulation, 2011.
41. J.R. Srinivasan, "Improving cache utilization," Technical Report, Univ. of Cambridge, 2011.
42. Advanced Configuration and Power Interface (ACPI) specification [online], Available: www. acpi.info.
43. Rotem, E. and Engineer, S.P., 2015, August. Intel Architecture, Code Name Skylake Deep Dive: A New Architecture to Manage Power Performance and Energy Efficiency. In Intel Developer Forum.
44. Rotem, E., Weiser, U.C., Mendelson, A., Ginosar, R., Weissmann, E. and Aizik, Y., 2016. H-EARtH: Heterogeneous Multicore Platform Energy Management. Computer, 49(10), pp. 47–55.
45. "Processor Package and Core C-States". AnandTech. 2013–06-09. Retrieved 2013-10-20.
46. Rotem, E., Naveh, A., Ananthakrishnan, A., Weissmann, E. and Rajwan, D., 2012. Power-management architecture of the intel microarchitecture code-named sandy bridge. Ieee micro, 32(2), pp. 20–27.

47. Mahesri, A. and Vardhan, V., 2004, December. Power consumption breakdown on a modern laptop. In International Workshop on Power-Aware Computer Systems (pp. 165–180). Springer, Berlin, Heidelberg.
48. David, Howard, Chris Fallin, Eugene Gorbatov, Ulf R. Hanebutte, and Onur Mutlu. "Memory power management via dynamic voltage/frequency scaling." In Proceedings of the 8th ACM international conference on Autonomic computing, pp. 31–40. ACM, 2011.
49. Rotem, E., Naveh, A., Rajwan, D., Ananthakrishnan, A. and Weissmann, E., 2011, August. Power management architecture of the 2nd generation Intel® Core microarchitecture, formerly codenamed Sandy Bridge. In Hot Chips 23 Symposium (HCS), 2011 IEEE (pp. 1–33). IEEE.
50. Jahagirdar, S., George, V., Sodhi, I. and Wells, R., 2012, August. Power management of the third generation Intel Core micro architecture formerly codenamed Ivy Bridge. In Hot Chips 24 Symposium (HCS), 2012 IEEE (pp. 1–49). IEEE.
51. Fayneh, E., Yuffe, M., Knoll, E., Zelikson, M., Abozaed, M., Talker, Y., Shmuely, Z. and Rahme, S.A., 2016, January. 4.1 14 nm 6th-generation Core processor SoC with low power consumption and improved performance. In Solid-State Circuits Conference (ISSCC), 2016 IEEE International (pp. 72–73). IEEE.
52. Howse, B. and Smith, R., 2015. Tick Tock On The Rocks: Intel Delays 10 nm, Adds 3rd Gen 14 nm Core Product KabyLake.
53. Intel. Intel® 64 and IA-32 Architectures Optimization Reference Manual, June 2016.
54. Nalamalpu, A., Kurd, N., Deval, A., Mozak, C., Douglas, J., Khanna, A., Paillet, F., Schrom, G. and Phelps, B., 2015, June. Broadwell: A family of IA 14 nm processors. In VLSI Circuits (VLSI Circuits), 2015 Symposium on (pp. C314-C315). IEEE.
55. E. Rotem, "Intel® Architecture, Code Name Skylake Deep Dive: A New Architecture to Manage Power Performance and Energy Efficiency," presented at the Intel Developer Forum (IDF15), 2015.
56. Isci, C., Martonosi, M.: Runtime power monitoring in high-end processors: Methodology and empirical data. In: Proceedings of the 36th Annual IEEE/ACM International Symposium on Microarchitecture, p. 93. IEEE Computer Society (2003).
57. Cai, G. and Lim, C.H., 1999. Architectural level power/performance optimization and dynamic power estimation. Cool Chips Tutorial colocated with MICRO32, 621.
58. D. Brooks, V. Tiwari, and M. Martonosi. Wattch: A Framework for Architectural-Level Power Analysis and Optimization. In Proceedings of the 27th International Symposium on Computer Architecture, pages 83–94, Vancouver, Canada, June 2000.
59. W. Wu, L. Jin, J. Wang, P. Liu, and S. X.-D. Tan. A systematic method for functional unit power estimation in microprocessors. In Proceedings of the 43rd Conference on Design Automation, July 2006.
60. J. Peddersen and S. Parameswaran. CLIPPER: counter-based low impact processor power estimation at run-time. In Asia and South Pacific Design Automation Conference, Jan. 2007.
61. C. Isci and M. Martonosi. Runtime power monitoring in high-end processors: Methodology and emprical data. In 36th International Symposium on Microarchitecture (MICRO 36), pages 93–104, Dec. 2003.
62. R. Joseph and M. Martonosi. Run-time power estimation in high-performance microprocessors. In International Symposium on Low Power Electronics and Design, pages 135–140, Aug. 2001.
63. J. Sharkey, A. Buyuktosunoglu, and P. Bose. Evaluating design tradeoffs in on-chip power management for cmps. In Proceedings of the International Symposium on Low Power Electronics and Design, pages 44–49, Aug. 2007.
64. Powell, Michael D., et al. "CAMP: A technique to estimate per-structure power at run-time using a few simple parameters." High Performance Computer Architecture, 2009. HPCA 2009. IEEE 15th International Symposium on. IEEE, 2009.
65. Bellosa, Frank. "The benefits of event: driven energy accounting in power-sensitive systems". In: Proceedings of the 9th Workshop on ACM SIGOPS European Workshop: Beyond the PC: New Challenges for the Operating System, pp. 37–42. ACM (2000)

66. Rotem, Efraim, Alon Naveh, Avinash Ananthakrishnan, Eliezer Weissmann, and Doron Rajwan. "Power-management architecture of the intel microarchitecture code-named sandy bridge." IEEE Micro 32.2 (2012): 0020–27.
67. B. C. Lee and D. M. Brooks. Accurate and efficient regression modeling for microarchitectural performance and power prediction. In Proceedings of the Twelfth International Conference on Architectural Support for Programming Languages and Operating Systems (ASPLOS XII), Oct. 2006.
68. P. J. Joseph, K. Vaswani, and M. J. Thazhuthaveetil. Construction and use of linear regression models for processor performance analysis. In Twelfth International Symposium on High Performance Computer Architecture (HPCA), pages 99–108, Feb. 2006.
69. E. Ipek, S. McKee, B. de Supinski, M. Schulz, and R. Caruana. Efficiently exploring architectural design spaces via predictive modeling. In Proceedings of the Twelfth International Conference on Architectural Support for Programming Languages and Operating Systems (ASPLOS XII), Oct. 2006.
70. E. Macii and M. Pedram. High-level power modeling, estimation, and optimization. IEEE Transactions on Computer-Aided Design of Integrated Circuits and Systems, 17(11):1061–1079, Nov. 1998.
71. S. Katkoori and R. Vemuri. Architectural power estimation based on behavior level profiling. Journal on VLSI Design, Special Issue on Low Power, 1996.
72. V. Srinivasan, D. Brooks, Michael Gschwind, P. Bose, V. Zyuban, P. N. Strenski, and P. G. Emma. Optimizing pipelines for power and performance. In Proceedings of the 35th International Symposium on Microarchitecture (MICRO 35), pages 333–344, Nov. 2002.
73. V. Zyuban and P. Strenski. Unified methodology for resolving power-performance tradeoffs at the microarchitectural and circuit levels. In Proceedings of the International Symposium on Low Power Electronics and Design, pages 166–171, Aug. 2002.
74. M. T.-C. Lee, V. Tiwari, S. Malik, and M. Fujita. Power analysis and minimization techniques for embedded dsp software. IEEE Transactions on VLSI Systems, 5(1):123–135, Mar. 1997.
75. P. Landman. High level power estimation. In Proceedings of the International Symposium on Low Power Electronics and Design, pages 29–35, Aug. 1996.
76. D. Brooks, V. Tiwari, and M. Martonosi. Wattch: A framework for architectural- level power analysis and optimizations. In Proceedings of the 27th Annual International Symposium on Computer Architecture, pages 83–94, June 2000.
77. D. M. Brooks, P. Bose, V. Srinivasan, M. K. Gschwind, P. G. Emma, and M. G. Rosenfield. New methodology for early-stage microarchitecture- level power-performance analysis of microprocessors. IBM Journal of Research and Development, 47(5/6):653–670, Sept. 2003.
78. T. S. Karkhanis and J. E. Smith. A first-order superscalar processor model. In Proceedings of the 31st International Symposium on Computer Architecture (ISCA 31), pages 338–349, June 2004.
79. D. B. Noonburg and J. P. Shen. Theoretical modeling of superscalar processor performance. In Proceedings of the 27th International Symposium on Microarchitecture (MICRO 27), pages 52–62, Nov. 1994.
80. E. Rotem, "Intel® Architecture, Code Name Skylake Deep Dive: A New Architecture to Manage Power Performance and Energy Efficiency," presented at the Intel Developer Forum (IDF15), 2015.
81. Gonzalez, Ricardo, and Mark Horowitz. "Energy dissipation in general purpose microprocessors," IEEE J. Solid-State Circuits, Vol. 31, No. 9, Sept. 1996, pp. 1277–1284.
82. Intel 64 and IA-32 Architectures Software Developer's Manual, Volume 3, Section 14.9 (as of November 2014).

Chapter 2
Dynamic Optimizations for Energy Efficiency

2.1 Introduction

Energy efficiency has become one of the most important design parameters for hardware, due to battery life on mobile devices and energy costs and power provisioning in data centers. Performance features like Dynamic Voltage and Frequency Scaling (DVFS), Turbo Boost or memory prefetching are offered by hardware manufacturers for software use. However, utilizing such features is tricky as it comes with a power cost. Completely disabling these features often incur significant slowdowns that may also waste battery budget. In some scenarios, performance is favored over power (e.g., responsiveness in smartphones). In other scenarios, a good enough performance can be quite effective at sustaining a long battery life (e.g., video playback).

In this work, we propose a novel *framework* called DOEE (Dynamic Optimization for Energy Efficiency) that optimizes the system energy efficiency. A *dynamic* approach is adopted where CPU energy telemetry and performance counters are periodically sampled. The framework is *adaptive* where it searches for an optimal point meeting a *user-supplied* metric for energy efficiency. Metrics like energy, performance, or combinations of them are illustrated in Sect. 2.

We *demonstrate* the framework using the Turbo Boost [1] feature available on modern Intel processors. The framework spares the power budget in order to enable Turbo in potential phases in more energy-efficient manner, resulting in reduced energy with nearly the same or better performance. The results shown outperform the energy efficiency algorithm implemented by the firmware of Intel processors. To the best of our knowledge, this is the first work that attempts to tune Turbo Boost.

To illustrate the idea, Fig. 2.1 shows the performance scalability over time of a sample application running with fixed frequency. Consider the phases A and B with average scalability of 97% and 62%, respectively (both are 10 s long). Scalability is

© Springer Nature Singapore Pte Ltd. 2018
J. Haj-Yahya et al., *Energy Efficient High Performance Processors*,
Computer Architecture and Design Methodologies,
https://doi.org/10.1007/978-981-10-8554-3_2

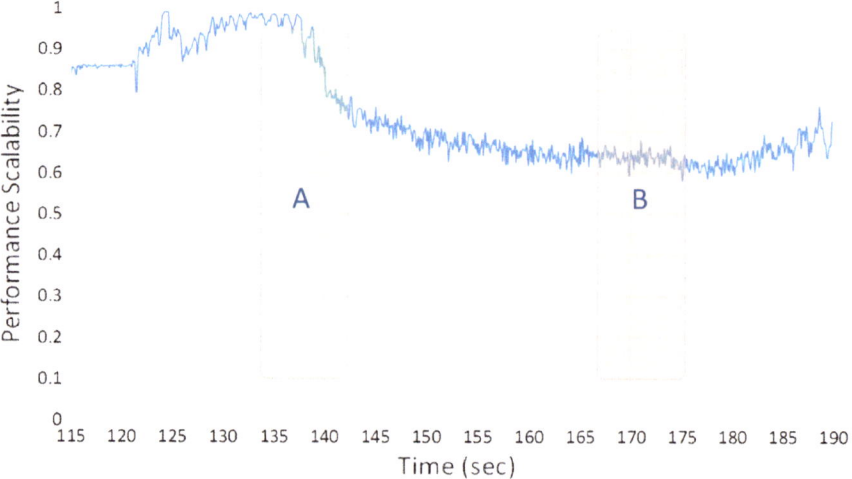

Fig. 2.1 Performance scalability over time for 429.mcf on Haswell machine (seconds 115–190)

Table 2.1 Energy efficiency comparison of the two phases

Phase	Average scalability (%)	Delay (s)	Energy (J)	EDP	ED^{2}P
A	97	7.8	140.7	1106.8	8704.31
B	62	8.5	150.4	1282.3	10926.7
Gain at A over B (%)		7.7	6.4	13.6	20.3

performance and frequency correlation; a ratio of 1 means doubling frequency results in doubling performance.

Table 2.1 describes the energy efficiency differences by applying the Turbo at A or B. We can see that by applying the Turbo at phase A instead of applying it at phase B, the overall runtime was reduced by 7.7% and an energy gain of 6.4% is achieved. Other metrics should even bigger improvements.

The main contributions of this work:

- A novel framework is developed for Dynamic Optimization for Energy Efficiency: DOEE is simple (no prior calibration is required), configurable (user metric) and adaptive (per dynamic characteristics).
- The DOEE framework is demonstrated by tuning the Turbo feature using the LLVM compiler. The proposed solution outperforms IvyBridge [19] processor built-in algorithm [17].
- The framework/implementation is made available for the research community [26]. Additionally, a few architectural enhancements are proposed to aid such approaches.

2.2 Background

2.2.1 Turbo

Turbo Boost [1]—also known as Intel Turbo Boost Technology 2.0 introduced with Intel's 2nd Generation Core™ processor—opportunistically boosts the frequencies of the cores in multi-core Intel Processors. The processor hardware controls the Turbo Boost activation and the level of boosting depends on the number of active cores, estimated power consumption, and the temperature of the package. This "thermal boosting" allows the processor to temporarily exceed the Thermal Design Point (TDP) using the thermal capacitance of the package. Figure 2.2, [12] illustrates the Turbo behaviors over time. In the first phase (sleep or low power), the processor gains thermal budget while sleeping or running at low power. In the second phase (C0/P0) the processor moves to Turbo (P0 in the P-state terminology [17]) following a P0 request by the software. In this stage, the processor starts to consume its thermal budget and at a later stage, as the processor heats up, the processor's hardware starts to reduce the frequency as its temperature gets close to the maximum allowed temperature. Once the entire thermal budget is consumed the processor's frequency normally stabilizes at the frequency corresponding to TDP frequency. The processor is not allowed to go back to Turbo until a new thermal budget is accumulated.

Turbo Boost might be energy inefficient in some cases. To cope with this problem, Intel's modern processors feature the ability for Software to control the energy efficiency of the processor by configuring the IA32_ENERGY_PERF_BIAS register [17]. The processor can be configured to favor highest performance, maximum energy savings, or a value in-between. Notice the energy efficiency problem is not straightforward to be solved by hardware alone. Thus, as

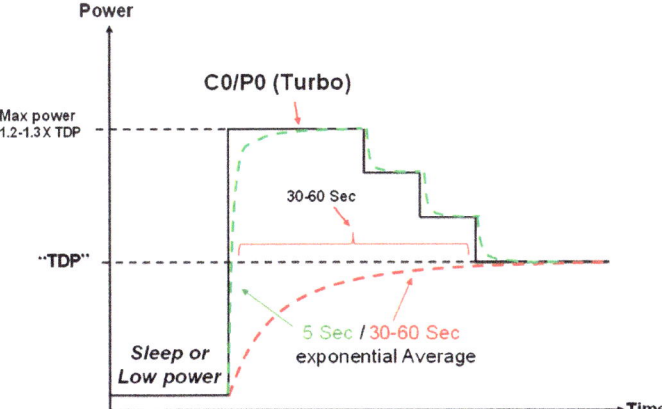

Fig. 2.2 Turbo boosts behavior overtime [12], and thermal budget gained at low-power intervals is used to temporarily run above TDP frequency

software-assisted approach is used to give hints to the underlying hardware about the software preferences to better handle energy efficiency. For example, different vendors might prefer different efficiency metrics. Commonly used ones are illustrated in the remainder of this section.

2.2.2   Energy Efficiency Metrics

The metric of interest in power studies varies depending on the goals of the work and the type of platform being studied. In some situations, focusing solely on energy is not enough. For example, reducing energy at the expense of lower performance may often not be acceptable. On the other hand, gaining performance at the expense of high energy consumption might not be practical for the system under design. Thus, metrics combining energy and performance have been proposed. Here we give a short survey of the various metrics used.

2.2.2.1   Energy

This metric is important in mobile systems. The unit of energy is Joules. Energy usage, which is closely correlated to battery life and battery capacity, is usually measured in watt-hours (Wh) which is an energy unit (1 Wh = 3600 J). Energy is also important in non-mobile platforms. For data centers [6], energy consumption is one of the leading operating costs (electricity bills), and thus reducing the energy usage is critical in these systems.

2.2.2.2   Power

Power is the rate at which energy is consumed. The unit of power is watts (W), which is Joules per second. This metric is important for designing the power delivery network (current and voltage requirements). In addition, this helps in understanding the power density of the system, which is used for thermal studies in the process of building a cost-efficient cooling solution.

2.2.2.3   Energy-Delay Product (EDP)

Is a metric that was proposed [8] to take into account the energy and performance at one metric. If either energy or delay increases, the EDP will increase. Thus, lower EDP values are desirable. EDP's inclusion of runtime means that this is a metric that improves with approaches that either hold energy constant but execute the same instruction mix faster, or hold performance constant but execute at a lower energy, or some combination of the two.

2.2.2.4 Energy-Delay-Squared Product (ED²P)

ED^2P [9–11] is similar to EDP but gives more weight to performance (1/delay) than energy cost.

2.3 Methodology

To improve the energy efficiency, we used an automatic search method for finding the optimum configuration for a given energy efficiency metric. For each program phase (loop or function) we do a competition, traversing possible configurations of the feature, subject to dynamic tuning. The algorithm is described in Fig. 2.3, we start the search from the highest allowed frequency (P0) down to the lowest allowed frequency (Pn) and stop the search once we reach a frequency that minimizes the energy efficiency metric. The metric itself is configurable by the user—such as Energy or EDP discussed in the previous section. The next runs of that program phase will use the best-performing configuration found by the search (competition) phase, optimal frequency in our example.

The instrumented code by the framework can be divided into a few stages. Stage I is the *search stage* (light blue color), where we try various configurations (e.g., various frequencies) and we choose the optimal one for the particular program phase. In stage II we apply the winning configuration chosen in stage I to future runs of the same program phase (green color). As the optimal frequency might change due to changes at the platform (e.g., number of cores running, display or imaging state, communication traffic [22]), there is a periodic research (waiting in stage III in orange color), after a periodic time we trigger the search stage again to capture system-level changes.

The algorithm is applied to program phases, which generally start at the function's entry and loop's entry (pre-header of the loop). These points are the start of potential program phases. At the first inspection of the potential phase, we capture the energy accumulator counter (RAPL [17]) and the Time Stamp Counter (TSC). At the next inspection, we capture the same counters and calculate the delta from the previous sample. Figure 2.4 describes the entry point location in case of loop or function. Both counter value captures are done at the entry point to the loop or the function, as this allows for the capturing of the execution of serial loops as shown in Fig. 2.4a. Moreover, recursive function calls will be captured at the entry of the function shown in Fig. 2.4b. To minimize the number of configuration change overhead, short (below some instruction threshold, e.g., 100 thousand instructions) function and loops were skipped. In such case instructions counter is not reset at the exit from the loop or function, but rather resume counting until we return back to the entry point of the specific function or loop. This method allows capturing long chains of short functions/loops that are repeatedly executed (e.g., in Fig. 2.4a: Inner_A → Inner_B → Inner_C → Inner_A).

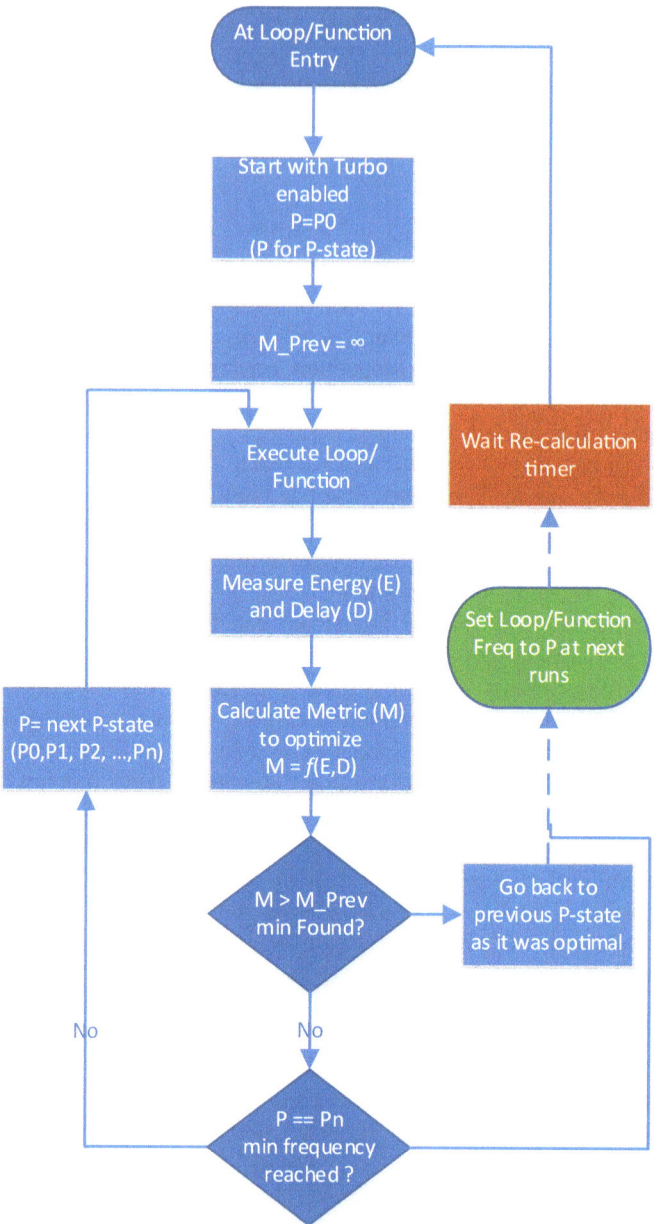

Fig. 2.3 Auto-tuning algorithm to find frequency point (P) that minimizes the evaluated metric (M)

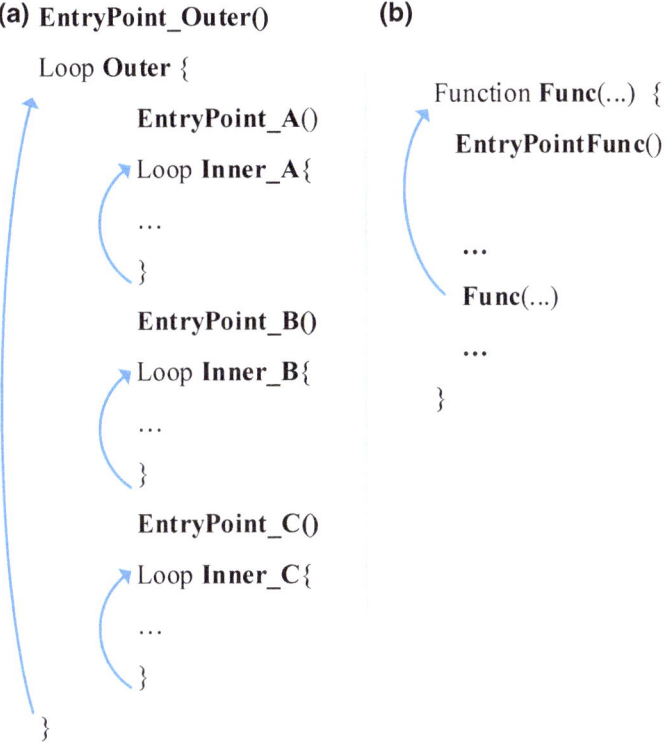

(a) EntryPoint_Outer()

Loop **Outer** {

        EntryPoint_A()

        Loop **Inner_A**{

          ...

        }

        EntryPoint_B()

        Loop **Inner_B**{

          ...

        }

        EntryPoint_C()

        Loop **Inner_C**{

          ...

        }

}

(b)

Function **Func(**...**)** {

    EntryPointFunc()

    ...

    Func(...**)**

    ...

}

Fig. 2.4 Entry points to loops and functions where the auto-tuning algorithm is invoked

2.4 Implementation

The framework consists of the LLVM (Low Level Virtual Machine) compiler [18], performance counters out of the Performance Monitoring Units (PMU) [17], DVFS, energy telemetry (Intel RAPL [17]). We have modeled the auto-tuning algorithm that uses the data to choose more energy-efficient work. In addition, we wrote a kernel-mode driver module to configure the Performance Counters at startup. The performance counters and TSC are queried from user land using the RDPMC instruction in order to limit the runtime overhead. The framework is described in Fig. 2.5.

The program is being compiled with a modified LLVM compiler. The compiler instruments auto-tuning code that implements an algorithm in Fig. 2.3, and outputs assembly code for the target machine.

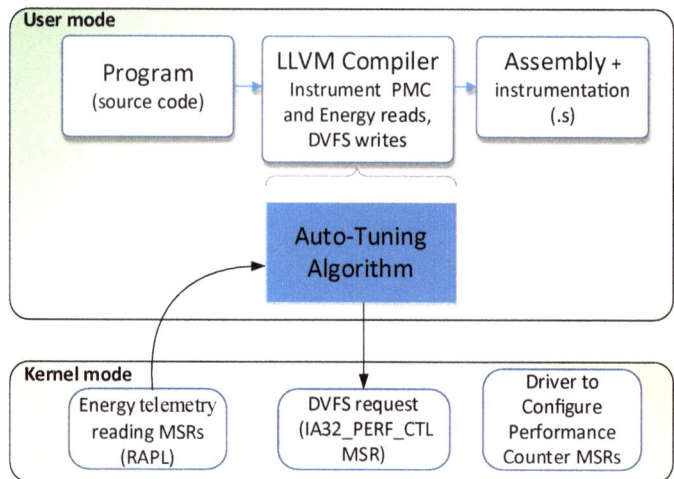

Fig. 2.5 Block-diagram of main framework's components

2.4.1 DVFS

In the Intel processors, DVFS is controlled by choosing Performance state (P-state), this is done by writing to per-thread Model-Specific Register (MSR) accessed by the kernel. The power management unit considers the requests from all running threads and sets the frequency to the maximum frequency requested by all of the threads. The requested P-states are values between P0 to Pn, where P0 is the maximum frequency with Turbo enabled, P1 is the maximum guaranteed frequency and Pn is the lowest available frequency. In the examined system, the P0 is 2.8 GHz, P1 is 1.9 GHz and Pn is 0.8 GHz. The DVFS request and the Turbo control is done by writing to the IA32_PERF_CTL MSR [17].

2.4.2 Energy Measurements

Intel introduced the Running Average Power Limit (RAPL) feature with the Sandy Bridge microarchitecture [12].

These energy counters [17] give visibility to the energy consumption of the package, cores, graphics (in client processor), and memory domains (in server processor). Today there is no option to read the whole platform energy consumption (energy consumed from the battery or power source); we expect that this option will be available in the near future by the processors vendors.

RAPL data can be configured and examined by reading MSRs. On the Intel architecture, today this is only possible in privileged kernel mode.

The available RAPL energy counter for cores domain is for all the cores together. Several studies have been done for CPU performance modeling per-core and per-thread energy using performance counter [13–16]. The PKG_ENERGY_ STATUS MSR that reflects the energy consumed by the whole CPU package was used in this study.

2.4.3   Code Instrumentation

The auto-tuning code is instrumented at the beginning of program phases (functions and loops), and in addition, the compiler adds global variables that enable tracking the frequency changes by programs phases (e.g., not setting the frequency to the same value twice).

2.4.4   Architectural Enhancements

We propose the following enhancements to aid DOEE-based optimizations and reduce overheads of the current method:

- Add new user level DVFS instruction to reduce the overhead of presented method.
- Faster DVFS transition implementation. We believe that this is feasible with the new technologies of on-die voltage regulators [21].
- Low latency RDPMC and RDMSR instructions, especially reading the RAPL energy MSRs.
- Higher resolution energy reporting. In this study, RAPL counters are updated nearly every 1 ms. More frequent updates enable finer-grain optimization.

2.5   Results

The implementation was tested on a 3rd Generation Intel® Core™ i7 3517U Processor code name Ivybridge. SPEC CPU2006 benchmarks [20] are measured in rate single-copy configuration using the reference input sets. Compiler flags used (−O3). For benchmarks with more than one input we used the first one. The IA32_ENERGY_PERF_BIAS MSR was set to 7. A value of 7 translates into a hint to the processor to balance performance with energy consumption while a value of 0 or 15 translated to highest performance and highest energy savings, respectively [17].

Figure 2.6 shows the results of the runs to improve the Turbo energy efficiency, four metrics are shown: Energy, Delay, EDP, and ED^2P. In addition, the graph shows the average Ext. Memory Bound [27] for each benchmark. This metric

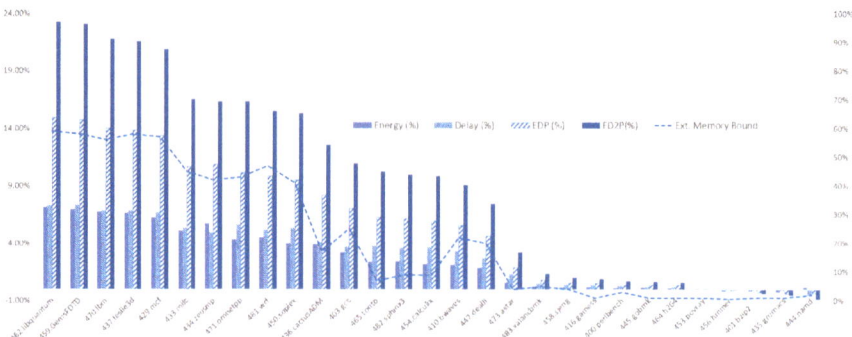

Fig. 2.6 Auto-tuning algorithm measured gains on CPU2006 workloads for commonly used metrics

represents the fraction of time where the processor's execution units are stalled due to memory accesses missing all caches. It is plotted just to validate the results (not used by the algorithm/implementation).

Almost all benchmarks show improvements over baseline considering all metrics. The benchmarks with biggest gains are those that are highly memory bounded (low performance scalability). For example, [27] reports that external memory is the primary bottleneck for 410.bwaves, 433.milc, 437.leslie3d, and 470.lbm. Note how the Ext. Memory Bound correlates with the gains. This makes sense as Turbo is not efficient in memory bound phases and it is rather spared for non-memory bound phases.

Benchmarks that are not memory bounded did not have improvements in the metrics as expected as all the programs phases are nearly the same and do not favor running with Turbo in one phase versus the other. As an example of compute bound, 456.hmmer has loops with tight data-dependent arithmetic instructions [27]. Some of these workloads even showed degradation in the metrics like 444.namd due to instrumentation overhead.

One of the parameters that we tuned manually was the number of instructions threshold on which we skip the handling of a phase, i.e., if at some phase (function or loop) the number of instructions executed is below this threshold then we skip the phase (do not instrument search for it).

Figure 2.7 shows the performance improvements of the metrics with different instructions skip threshold for one benchmark. The graph shows that low thresholds result in a loss in all metrics; this is expected as such thresholds result in small program phases being instrumented with the auto-tune code, which would be of high overhead. Best results are achieved near threshold of 100 K instructions. It has a good balance between instrumentation overhead and program-phase size leading in a high net metrics gain. At a threshold of 1 M we see that some metrics have small improvements while others have loses; this occurs since fewer phases are handled (most phases are skipped) while the initial overhead of measuring the long

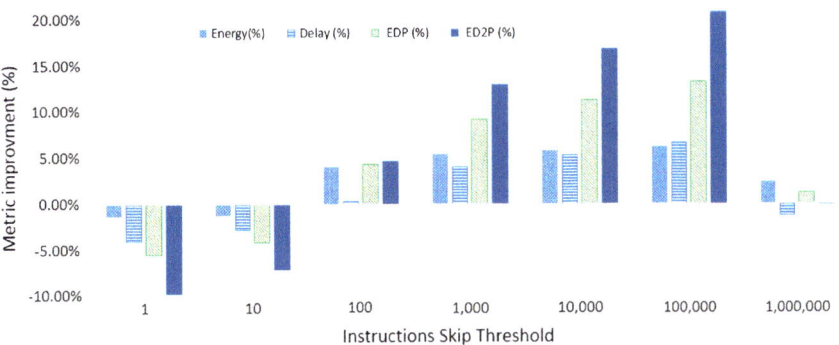

Fig. 2.7 Per-metric gains as a function of skip phase instructions threshold for 429.mcf

phases contributed to the metrics' losses. Note there is high chance for long phases to combine functions/loops with different characteristics.

The auto-tuning algorithm uses the highest granularity of frequency steps supported by the processor (the difference between P_i and P_{i+1} P-states is 100 MHz). This results in high search overhead at the *search phase* in some cases, as the algorithm traverses over the whole range from P_0 to P_n (12 options in our case-from 1900 MHz down to 800 MHz). Figure 2.8 shows the ED^2P metric while using few frequency steps (steps of 100, 200, 300, and 400 MHz are used). For the less-scalable benchmarks, e.g., 462.libquantum, the gain was reduced with bigger steps; this is mainly due to the fact that a nonoptimal frequency was chosen. For middle benchmarks, changing the steps did not affect the gain much as in case of 436.cactusADM (at 100 and 200 MHz). For most-scalable benchmarks, e.g., 435.gromacs, the ED^2P metric loses with 100 MHz step, while we start to see small gains when bigger steps are used. This can be explained as the search overhead was reduced significantly with bigger frequency steps.

The results in Fig. 2.6 are compared to baseline of the balanced performance with energy consumption (setting IA32_ENERGY_PERF_BIAS MSR to value 7) configuration of Intel's energy efficiency algorithm. Such configuration is relevant to metrics that combine performance (1/delay) with energy like EDP or ED^2P.

In Fig. 2.9, we show the various metrics gains with three configurations of Intel's energy efficiency algorithm for 462.libquantum benchmark. The configurations are performance, balanced, and energy saver with IA32_ENERGY_PERF_BIAS MSR values of 0, 7, and 15, respectively. Figure 2.9 shows that at the performance configuration, the framework has the height gain at the energy metric versus the other configurations. While at the energy saver configuration, the framework has the highest savings at the delay metric. At the balanced configuration, the metrics of EDP and ED^2P are highest among the three configurations as expected.

To enhance the algorithm and reduce its overhead the following ideas can be explored in a future work:

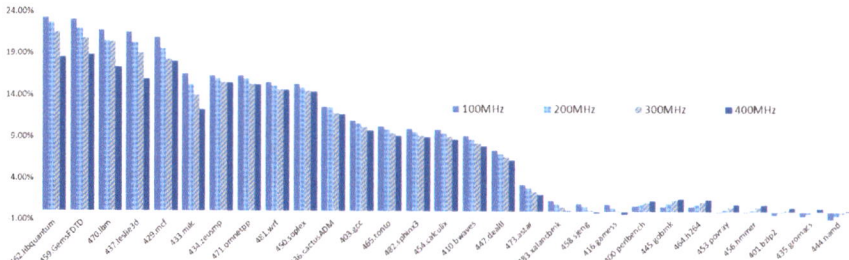

Fig. 2.8 ED2P metric improvement with 100, 200, 300, and 400 MHz frequency search steps

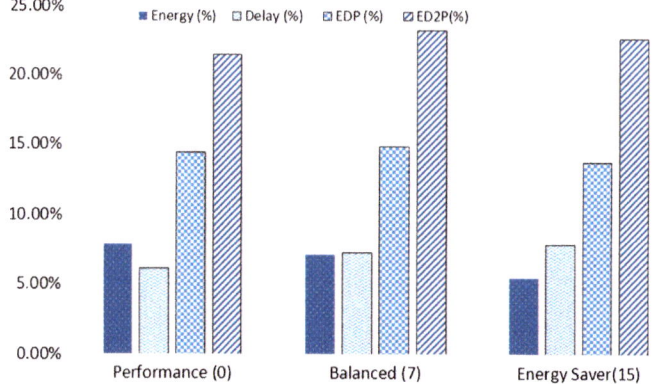

Fig. 2.9 Metrics gains at different policies of the Intel's energy efficiency algorithm for 462.
libquantum benchmark

- Do a binary search to find the frequency work point that will minimize the
  energy efficiency metric.
- Execute a fixed number of search steps, e.g., try just the first five frequency
  options starting from the default one and pick the one that minimizes the energy
  efficiency metric along the checked options.
- We have demonstrated less-granular tuning in the search phase can reduce
  overhead (while it might not choose the best-performing frequency for the
  particular benchmark). Further adaptation seems promising, e.g., to choose the
  step size dynamically based on current phase characteristics like code size,
  execution time memory boundness, and so on.
- Current implementation skips functions with small size after passing first
  function's execution. This is done in case the function has low instructions
  count. This process can be optimized by doing it at compile time, as the code
  size can be estimated at compile time in some conditions. This should be done
  while taking into account calls to other functions from within the original
  function or loops that exist at the function body. For functions that are marked
  as short, we will not instrument auto-tuning code to them, and this will save
  code size and latency.

2.6 Related Works

Although some works [2–4] point out that the DVFS gain is reducing with new process technologies, we do see good energy efficiency gain by applying DVFS. This gain is further increased by Turbo which facilitates high voltage and power that exceed the processor's TDP. Despite its high-power overhead, Turbo boosts performance [5].

Previous works have used dynamic methods to improve energy efficiency; Wu et al. [23] developed a dynamic compilation method using Just-In-Time (JIT) compiler that adds code to potential program phases (functions and loop) and activates DVFS as a function of memory boundness of each phase. This method has high performance and energy overhead as it activates JIT in runtime to optimize every phase, and requires manual tuning of the memory boundness thresholds. In comparison, the proposed method has lower overhead as the compilation is done statically while the auto-tuning code figures out the tuning parameters dynamically. Moreover, the method takes into account the platform's dynamic changes as opposed to looking just at the memory boundness which is less accurate with respect to system-level optimization.

Koukos et al. [3] proposed a dynamic method for separating memory access phases and execution phases, where low frequency is applied for access phases and high frequency is applied for execution phases. The technique has high overhead since it requires running each phase twice: the first run with only memory accesses and associated address calculations to prefetch the data of that phase to caches, whereas the second run has both memory and execution instructions. The evaluation was done through a simulator, an assumption that DVFS transitions overhead is near zero was made, and the separation of access and execute phases was done manually. Jimborean et al. [4] presented a compiler-based method to separate access and execute phases. The method searches for long-enough program phases (skip short ones) to which it applies the auto-tuning algorithm while considering platform changes. The evaluation uses measurements on production systems and does not assume zero overhead for DVFS transition or Energy counters query. Nonetheless, low overhead would likely increase the gain and will give better opportunities for tuning short phases.

Sasaki et al. [24] used other hardware performance information available to the operating system to make frequency change decisions. Their DVFS algorithm is based on statistical analysis of performance counters. By predicting the performance, the processor selects the lowest possible frequency that can maintain the performance degradation to a specified ratio. Their technique requires compiler support to insert code for performance prediction, static analysis, and per-platform tuning (to build the performance model). In comparison, the method has an auto-tuning algorithm that takes into account dynamic actual platform and workload characteristics with no need for per-platform calibration.

Another approach to dynamically set DVFS performance levels is to use a Performance Monitoring Unit (PMU) to detect when it is possible to achieve sublinear performance degradation. Isci et al. [25] use phase categories calculated using a metric for memory operations per micro-operation. Each phase category is linked to a DVFS policy that attempts to minimize EDP. This approach requires per-platform tuning and does not take into account package level energy.

Rotem et al. [22] presented an algorithm that finds an optimal voltage and frequency operational point of the processor in order to achieve minimum energy for the computing platform. The calibration is (again) per-platform and based on static profiling data, which was also used to validate the algorithm using a fixed power model. The method has an auto-tuning algorithm and is more comprehensive where it can optimize any user-supplied metric (not restricted to Energy).

The available Running Average Power Limit (RAPL [17]) energy counters account for all the cores together. There is no option readily accessible that allows reading the counters per-core or per-thread, although we believe that this option will be available in the future. Several studies have been done for CPU performance modeling per-core and per-thread energy using performance counters; Bellsoa's work [13] shows the linear correlation of hardware events and energy. Singh et al. [14] achieved a runtime per-core power estimation of multithread and multi-program workloads using the top-down method [15]. They categorize the processor's hardware events into four classes (because their environment platform has only four performance counters). Isci and Martonosi [16] decompose CPU into 22 power breakdowns based on functional units, which is a typical bottom-up approach [15]. Following that, they present a per-unit power estimation devised from performance counters.

2.7 Conclusions and Future Work

In this work, DOEE was developed—a novel method that optimizes processor features for energy efficiency using user-supplied metrics. The optimization is dynamic considering the runtime characteristics of the workload and the platform. We demonstrate that energy efficiency optimization is a challenging problem that is hard to solve using hardware-only methods; software hints are essential for accurate optimizations. The evaluation suggests that the method used outperforms Intel's energy-efficient algorithm implemented by the processors firmware.

We believe that future architectures will exploit software–hardware codesign to raise energy efficiency of computing systems. We hope to see enhancements at the software–hardware interface, such as DVFS control and the processor's telemetry reading which would enable further enhancements to dynamic optimization algorithms.

Even though the presented auto-tuning (simple) algorithm showed significant savings at the various energy efficiency metrics, further gains seem possible. The algorithm might have high overhead in cases that the function or loop block will be executed a few times, or in case that we have many options for the search. In the current evaluation, the processor frequency range is 800–1900 MHz where each frequency bin is 100 MHz (12 options in total to search). In other processors the range might be higher and more options will exist which would likely add more overhead to the exploration stage. In addition, the current framework does not handle multithreading and multi-core interferences due to contradicting frequency requests of the various threads, for example, when the auto-tuning decided to go to high frequency at one core while the opposite was chosen at the other thread, today the CPU will take the maximum request and raise the CPU frequency for both which might not be the most energy efficient at some metric. Future work should address these issues and enhance the presented framework.

References

1. Intel® Corporation. Intel® Turbo Boost Technology in Intel® Core™ Microarchitecture (Nehalem) Based Processors. Whitepaper, Intel® Corporation, November 2008.
2. Le Sueur, Etienne, and Gernot Heiser. "Dynamic voltage and frequency scaling: The laws of diminishing returns." Proceedings of the 2010 international conference on Power aware computing and systems. USENIX Association, 2010.
3. Koukos, Konstantinos, et al. "Towards more efficient execution: a decoupled access-execute approach." Proceedings of the 27th international ACM conference on International conference on supercomputing. ACM, 2013.
4. Jimborean, Alexandra, et al. "Fix the code. Don't tweak the hardware: A new compiler approach to Voltage-Frequency scaling." Proceedings of Annual IEEE/ACM International Symposium on Code Generation and Optimization. ACM, 2014.
5. Charles, James, et al. "Evaluation of the Intel® Core™ i7 Turbo Boost feature." Workload Characterization, 2009. IISWC 2009. IEEE International Symposium on. IEEE, 2009.
6. D. Dunn, "The best and worst cities for data centers," Information Week, Oct. 23, 2006 edition.
7. Grochowski, Ed, and Murali Annavaram. "Energy per instruction trends in Intel microprocessors." Technology@ Intel Magazine 4.3 (2006): 1–8.
8. R. Gonzalez and M. Horowitz, "Energy dissipation in general purpose microprocessors," IEEE J. Solid-State Circuits, Vol. 31, No. 9, Sept. 1996, pp. 1277–1284.
9. Zyuban, Victor, et al. "Integrated analysis of power and performance for pipelined microprocessors." Computers, IEEE Transactions on 53.8 (2004): 1004–1016.
10. Brooks, David M., et al. "Power-aware microarchitecture: Design and modeling challenges for next-generation microprocessors." Micro, IEEE 20.6 (2000): 26–44.
11. Flynn, M., Patrick Hung, and Kevin W. Rudd. "Deep submicron microprocessor design issues." Micro, IEEE 19.4 (1999): 11–22.
12. Rotem, Efraim, et al. "Power-management architecture of the intel microarchitecture code-named sandy bridge." IEEE Micro 32.2 (2012): 0020–27.
13. Bellosa, F.: The benefits of event: driven energy accounting in power-sensitive systems. In: Proceedings of the 9th Workshop on ACM SIGOPS European Workshop: Beyond the PC: New Challenges for the Operating System, pp. 37–42. ACM (2000).
14. Singh, K., Bhadauria, M., McKee, S.A.: Real time power estimation and thread scheduling via performance counters. ACM SIGARCH Computer Architecture News 37(2), 46–55 (2009).
15. Bertran, R., Gonzàlez, M., Martorell, X., et al.: Counter-Based Power Modeling Methods: Top-Down vs. Bottom-Up. The Computer Journal 56(2), 198–213 (2013).

16. Isci, C., Martonosi, M.: Runtime power monitoring in high-end processors: Methodology and empirical data. In: Proceedings of the 36th Annual IEEE/ACM International Symposium on Microarchitecture, p. 93. IEEE Computer Society (2003).
17. Intel 64 and IA-32 Architectures Software Developer's Manual, Volume 3, Section 14.9 (as of November 2014).
18. Lattner, Chris, and Vikram Adve. "LLVM: A compilation framework for lifelong program analysis & transformation." Code Generation and Optimization, 2004. CGO 2004. International Symposium on. IEEE, 2004.
19. James, Dick. "Intel Ivy Bridge unveiled—The first commercial tri-gate, high-k, metal-gate CPU." Custom Integrated Circuits Conference (CICC), 2012 IEEE. IEEE, 2012.
20. Standard Performance Evaluation Corporation, [online], Available: www.spec.org/.
21. Jain, Tarush, and Tanmay Agrawal. "The Haswell Microarchitecture-4th Generation Processor".
22. Rotem, Efraim, et al. "Energy Aware Race to Halt: A Down to EARtH Approach for Platform Energy Management." (2012): 1–1.
23. Wu, Qiang, et al. "A dynamic compilation framework for controlling microprocessor energy and performance." Proceedings of the 38th annual IEEE/ACM International Symposium on Microarchitecture. IEEE Computer Society, 2005.
24. Sasaki, Hiroshi, et al. "An intra-task dvfs technique based on statistical analysis of hardware events." Proceedings of the 4th international conference on computing frontiers. ACM, 2007.
25. Isci, Canturk, Gilberto Contreras, and Margaret Martonosi. "Live, runtime phase monitoring and prediction on real systems with application to dynamic power management." Proceedings of the 39th Annual IEEE/ACM International Symposium on Microarchitecture. IEEE Computer Society, 2006.
26. Jawad Haj-Yihia, LLVM compiler tool for raising energy efficiency, 2014, from Haifa University: https://drive.google.com/open?id=0B3IgzCqRS5Q_Yi1PbFZCTHpiMEU &authuser=0.
27. A. Yasin, "A Top-Down Method for Performance Analysis and Counters Architecture," in Performance Analysis of Systems and Software (ISPASS), IEEE International Symposium on, 2014.

Chapter 3
Power Modeling at High-Performance Computing Processors

3.1 Introduction

Power consumption has become a critical issue that affects the further development of processors of all computational segments since it impacts the battery life of battery-operated devices, power-related expenses dominate the cost of operation of data centers and since saving of power can be used for increasing the operational frequency, power is important for performance-driven systems as well. High power consumption adds challenges to the system design due to the effects of high thermal output, high current requirements, and battery life and electricity costs. In order to raise the awareness, Green500 [1, 2] lists the most energy-efficient supercomputers twice a year according to the ratio of performance and power consumption (as opposed to pure performance ranking). Apart from steering toward the best computational performance, much more attention has been drawn to data intensiveness and energy awareness, both of which are playing increasingly important roles to influence the design of future high-performance-computing systems and data centers. The Graph500 [1] is a widely used benchmark for rating supercomputer systems for data-intensive workloads. Graphs are a core part of most data analytics. Compared with compute-intensive benchmarks like HPL (High-Performance Linpack) which the Top500 List is based on, the Graph500 models complex data problems by performing Breadth-First Search (BFS) on a large-scale graph. Apart from the data-intensiveness aspect, energy efficiency is another vital design constraint on HPC systems and data centers. From a "green computing" perspective, energy consumption is as important as performance [3]. A new definition of supercomputer ranking is one that sorts computer systems by performance per watt for data-intensive workloads.

Furthermore, quantifying the power consumption of individual applications is a critical component for software-based power capping [4], scheduling, and provisioning techniques in modern datacenters. Power and energy consumption plays an important role in the maintenance cost of a large data center, including power

© Springer Nature Singapore Pte Ltd. 2018
J. Haj-Yahya et al., *Energy Efficient High Performance Processors*,
Computer Architecture and Design Methodologies,
https://doi.org/10.1007/978-981-10-8554-3_3

consumption from processors, memories, storage, and networking resources, and also from other infrastructures. Infrastructure providers decide their pricing policies based on the costs of the infrastructure, and charge customers based on fixed costs ratios as well as following the usage that they do of the computing resources. Therefore, detailed measurement of resource consumption is a critical point for shared data centers. Accounting of resource utilization must be done on a Per-Virtual Machine (VM) basis, even when multiple VMs are deployed on top of the same physical hardware. While measuring CPU, storage, and network utilization are feasible with existing system tools, per-VM energy consumption can be hardly estimated in resource-sharing environments. Power metering devices measure aggregate power consumption metrics for a whole system, but detailed per-VM metrics can only be measured with advanced techniques. The new many-core systems will take this limitation even further, envisioning environments in which a large number of VMs will be deployed on top of tenths of cores in a single server.

Over the past few decades, gains in energy efficiency came primarily from improvements in process technology. Each new process generation provided smaller transistors, which resulted in faster switching speeds and also lower power, due to proportional voltage reductions. Taken together, this resulted in an almost constant power density, as Dennard predicted [1]. However, threshold voltage scaling limitations and an increasing fraction of leakage power have ended this trend. Future process generations are not expected to deliver significant energy efficiency by themselves. This change places the burden of improving energy efficiency into the hands of the architects, who have to figure out how to use the transistors more efficiently.

Given the importance of power, it is critical for system designers and software developers to be able to observe, estimate, and analyze power consumption. Power estimation is fundamental for optimal scheduling in operating systems and for the development of power-efficient mechanisms in processor firmware. Therefore, it has recently attracted many researchers and yielded different important reference guides. Prior research indicates that power consumption can be estimated via monitoring performance counters, such as cycles, instructions, or cache accesses. These works proposed how to model power at coarse levels such as processor package or processor core mostly in simulations and some in real systems too. They can be categorized into top-down or bottom-up approaches. We summarize these in the related works section and provide a table with high-level comparison.

There are several techniques—working at different levels—addressing power consumption issues. All of them rely on accurate methods to gather information about power consumption. Specifically, for software-based solutions, the need to estimate and predict power behavior has justified the research on power-modeling strategies. Methods based on Performance-Monitoring Counters (PMCs) have been shown to be a good solution to estimate power consumption. As a result, their applicability has been demonstrated in several fields such as power management and application profiling. They are used to perform live predictions of power behavior in order to guide power-aware policies [5, 6]. Moreover, they are also used in research for quickly exploring new approaches since they allow to profile real systems and full executions of applications, avoiding the need to perform long-time and limited simulations [7, 8]. In the end, they have been crucial in the process of addressing power issues.

3.1.1   Problem and Scope

So far most of the software optimization guidelines do not pay extra attention to energy consumption and often assume that optimizing software for performance is the same as optimizing the code for energy [9, 10]. This assumption is not always true. For example, some implementations may activate the entire vector unit regarding the number of actual elements needs to be operated. Thus, although applying vector operations on short vectors may be efficient, in terms of performance, it may not be as efficient in terms of power. Another example can be found in [11, 12], where the authors compare between the "Race to Halt" schedule—that suggests running the system at maximum speed to be able to put the system in a sleep state for the longest time possible—and the Lowest-Frequency-Mode mechanism that calls to run the system at the lowest frequency possible so that the active power will be minimized. The work concludes that different systems may have different optimization points to minimize energy consumption. Such an energy optimization point mainly depends on the ratios between the system power, leakage power, and the dynamic power the system consumes. Finding the right mechanism that could bring the system to the best energy optimization point, usually involves a deep understanding of the power breakdown of the different microarchitectural components of the system.

Sometimes we need to combine different considerations as well; suppose we decide to impose the race-to-halt policy, but the maximum frequency the system can run needs to take into consideration that no single execution FUB (Functional Unit Block) will cause a thermal hotspot. In this case, distributing the work among different execution units; e.g., Floating-Point (FP) scalar operations and SIMD/vector operations will reduce the thermal hotspot but may also cause an increase of the overall energy due to extra time the system is now active. Thus, the optimal operatinal point for energy is not clear in such a case and we need new tools and power breakdown of components in order to determine what optimizations to embrace for each scenario.

The scope of this work focuses on CPU-bound workloads and on single-threaded core processors. CPU bound implies the workload does not include much CPU idleness (C-states [9]). In addition, our tool assumes that the CPU performance point (P-states [9]) and temperature do not change in the middle of the benchmarks runs.

3.1.2 Our Approach

We have developed a tool that accurately breaks down the power consumption of modern processors at sub-processor core granularity on real systems.

Our work reports the actual power (in watts) of the Intel's state-of-the-art core microarchitecture at abstracted domains: front-end portion (responsible for instructions/ fetch operations), execution and out-of-order engine and memory subsystem. We further model power consumption at processor core level and manage to achieve accuracy levels of package-level energy counters provided by the hardware vendors.

3.1.3 Contributions

This study makes the following contributions:

- We describe a methodology for modeling per-domain power consumption within out-of-order cores of modern (real) systems (Sect. 3.3) and provide a high-level comparison of difference among prior methods (survey) [Related Works].

  - The method relies on a select list of performance counters that were chosen for an empirical reason. We also identify how the specific performance counters correlate to the power consumption of each sub-domain.
  - The tool [35] shows up to 7% error rate relative to Intel RAPL [13] power reporting.

- We demonstrate/validate the method on Intel's 6th Generation Core (codename Skylake) using designated micro-benchmarks (Sect. 3.5.2), where we study their power and performance breakdowns and how it varies with common software optimizations such as loop unrolling or vectorization.
- We characterize power consumption of the SPEC CPU2006 benchmarks on Skylake (Sects. 3.5.3 and 3.5.4) study few interesting cases (Sect. 3.5.5), and include detailed power performance report [Appendix].
- We developed a tool that models the power consumption of Intel's Core™ and tuned it for the Skylake microarchitecture. The tool presents overall core power, per-domain breakdown, and the ability for over-time estimation (Sect. 3.4). We make this tool available to the research community as an open source [14].

3.2 Background

This section provides necessary power, microarchitecture, and performance analysis background used later in the methodology and results sections. This section does not intend to teach all these topics, but to set a common language we can use through the rest of the book

3.2.1 Power

Power is the rate at which energy is consumed; the unit of power is watts (W), which is joules per second. Power consumption in CMOS circuits can be described by a dynamic and a static part. The static part is mainly due to leakage, and the dynamic part can be further divided into switching power and short-circuit power. Switching power has the largest share of power consumption in CMOS circuits when the circuit is active, switching power can be expressed as shown in Eq. 3.1.

Equation 3.1: Power consumption in CMOS circuits

$$P = P_{dyn} + P_{sta} = P_{swi} + P_{sc} + P_{leak}$$
$$P_{swi} = \alpha C_L V^2 f$$
(3.1)

where V is the supply voltage, C_L are load and parasitic capacitances that are charged and discharged with frequency f when the corresponding component is active [15]. The switching activity α is highly dependent on the applications running and their input data.

3.2.2 Processor Energy Measurements

Intel introduced the Running Average Power Limit (RAPL [13]) feature with the Sandy Bridge microarchitecture [11, 12]. RAPL reports the energy consumption of the system components listed in Table 3.1.

These available counters enable reading the energy consumption of the package, cores, graphics (at client processor), and memory domains (at server processor). In Skylake, a new capability was added to read the whole platform energy consumption (energy consumed from the battery or power source). RAPL data can be configured and examined by reading Model-Specific Registers (MSRs) that require privileged kernel mode access today.

The PP0_ENERGY counter accounts for all cores within a multi-core processor combined with the Last Level Cache (LLC) to which these cores are connected. It is the closest publicly available reference to the granularities our work models.

To figure out the *average power* consumed by some benchmark at one of the processors domains, assuming no other programs are running except our benchmark, then we can sample the relevant RAPL counter (from Table 3.1) before and after executing the benchmark; the difference between the two samples gives the energy consumed by the benchmark at the relevant domain, dividing the energy by the execution time gives us the average power consumed by the benchmarks as *Energy = average power • execution_time*.

Table 3.1 List of available energy telemetries for Intel processors

Domain	Description
PKG_ENERGY	Energy of the whole processor package
PP0_ENERGY	Energy of power plane 0 domain (Processor IA Cores domain and Last Level Cache—LLC)
PP1_ENERGY	Energy of power plane 1 domain (specific to client processors where it is the integrated processor graphics)
DRAM_ENERGY	Energy of the DRAM domain (specific to server processors)
PLATFORM_ENERGY	Energy of the entire platform including IA Core, integrated graphics and the system agent, and other platform components. (new in Skylake)

3.2.3   Core Microarchitecture

At a high level, the pipeline of a modern out-of-order processor can be divided into three main portions: frontend, memory subsystem, and a backend. The frontend is responsible for fetching instructions from memory and "feed them" to the backend. Most of the modern architectures also distinguish between the architecture ISA which is exposed to the user and the tools (e.g., compiler) and the micro-operational ISA—a.k.a. micro-operations (uops) which are exposed only to the hardware. In such architectures, the front-end also is in charge for translating the code between the different ISAs. The backend is responsible for scheduling, executing, and retiring (commit) these uops per the original program's order. The backend can be further divided into execution and memory domains. The execution domain includes the out-of-order scheduler and the execution units while the memory domain includes memory subsystem portions, private to the core including the data cache and second-level cache, as depicted in Fig. 3.1.

From the optimization point of view, it is important to distinguish between the different root-causes of performance and power inefficiencies, since usually a program is primarily limited by only of these resources; e.g., the program can be memory bounded or CPU bounded. Understanding the root-causes is the key for future optimization of the system

3.2.4   Counter-Based Power Estimation

Recently many studies have been carried out on runtime power monitoring using PMCs. PMCs are available in various processors for performance measurement, but have been successfully used for power estimation. Bellosa et al. [16] discovered a strong correlation between system performance events, such as floating-point operations or cache misses, and required energy. The dynamic power consumption of a system with n PMCs can then be described by multiplying the performance

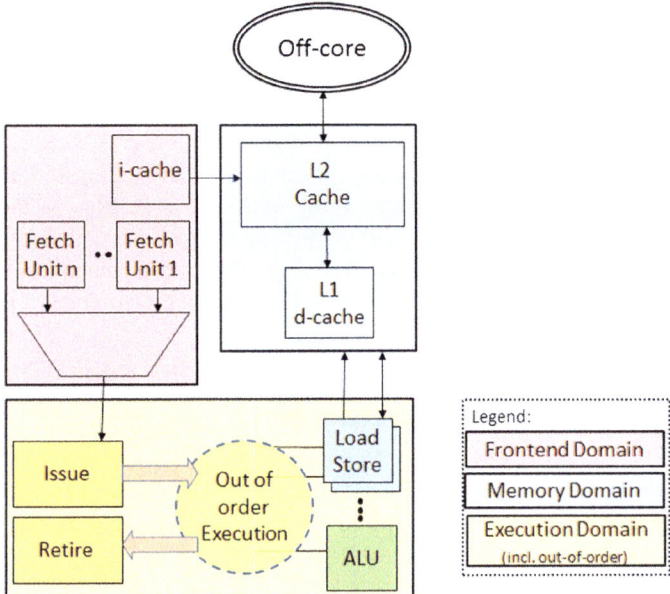

Fig. 3.1 Abstraction of a modern processor core microarchitecture

counter readings PMC$_i$ with the corresponding energy per performance event e_i divided by the sampling interval. Previous works such as [10, 17–19] estimate the power of core subunits by using a few counters, the work was done for a relatively old processor simulator (not real system) and their scheme does not report actual power or energy but pseudo-Joule metric.

3.2.5 Methods for Performance Analysis

We used the Top-down Microarchitecture Analysis Method (TMAM) [9, 20] for performance analysis. TMAM simplifies cycle-accounting (the process of identifying costs of performance bottlenecks, also often called CPI breakdown) for out-of-order cores using microarchitecture-independent metrics organized in one simple hierarchy.

At the top level, TMAM employs a single point of division where issue-pipeline slots are divided into four main categories: frontend bound, backend bound, bad speculation, and retiring. The latter two denote non-stalled slots while the former two denote stalls. A simple decision tree is used: if a slot is utilized by some operation, it would be classified as retiring or bad speculation, depending on whether it eventually gets retired. Unutilized slots are classified to backend bound if the back-end portion of the pipeline is unable to accept more operations (also known as backend stall [20]), or frontend bound where no operations are delivered while there was no backend stall.

The method further divides backend bound into memory bound or core bound in the second hierarchy level, depending on whether the backend-stalls are due to memory or non-memory operations, respectively. In the results section, we show memory bound and core bound instead of backend bound alongside the other 3 top-level categories. This is possible with the Skylake version of the TMAM method, where first and second hierarchy levels adopt the same counting domain in TMAM_Metrics electronic files posted at [21]. This representation is chosen in order to ease the comparison with our own power breakdown model.

3.3   Power Breakdown Methodology

For modeling the power of the CPU core and its sub-domains, we have used a linear function of performance counters while for each performance counter we have associated a weight as shown in Eq. 3.2.

Equation 3.2: Core and sub-domains' power estimation using performance counters

$$Power = Leakage + \sum_{i=0}^{n} \frac{\text{PerfCounter}_i}{time} \cdot \text{Weight}_i \qquad (3.2)$$

To determine the values of the weights, we used a set of representative micro-benchmarks that stress and train each part of the core domains. These micro-benchmarks were run both on the simulator and a real platform. By running a micro-benchmark on the simulator we got the power for the core and its sub-domain for the specific micro-benchmark. And by running the micro-benchmark on the real system, we were able to extract the values of the performance counters that were generated for the specific micro-benchmark. Both the power and the counters' data were fed into Eq. 3.2 in order to extract the values of the "weights" values after solving the linear regression. The flow for generating the power model is described in Fig. 3.2.

In this work, we will build a tool that estimates the power of the core and its sub-domains (frontend, execution, and memory). It is important to note that there is a core power component that is not part of these sub-domains. This power consists of the power of global components (P_{Global}) such as clock-grid, power gates' leakage, and others, this power is constant (at a given voltage and temperature), so the power of the core (P_{Core}) is equal to the sum of the power of the sub-domains— frontend (P_{FE}), execution (P_{EXE}), memory (P_{MEM})—and the power of global components (P_{Global}).

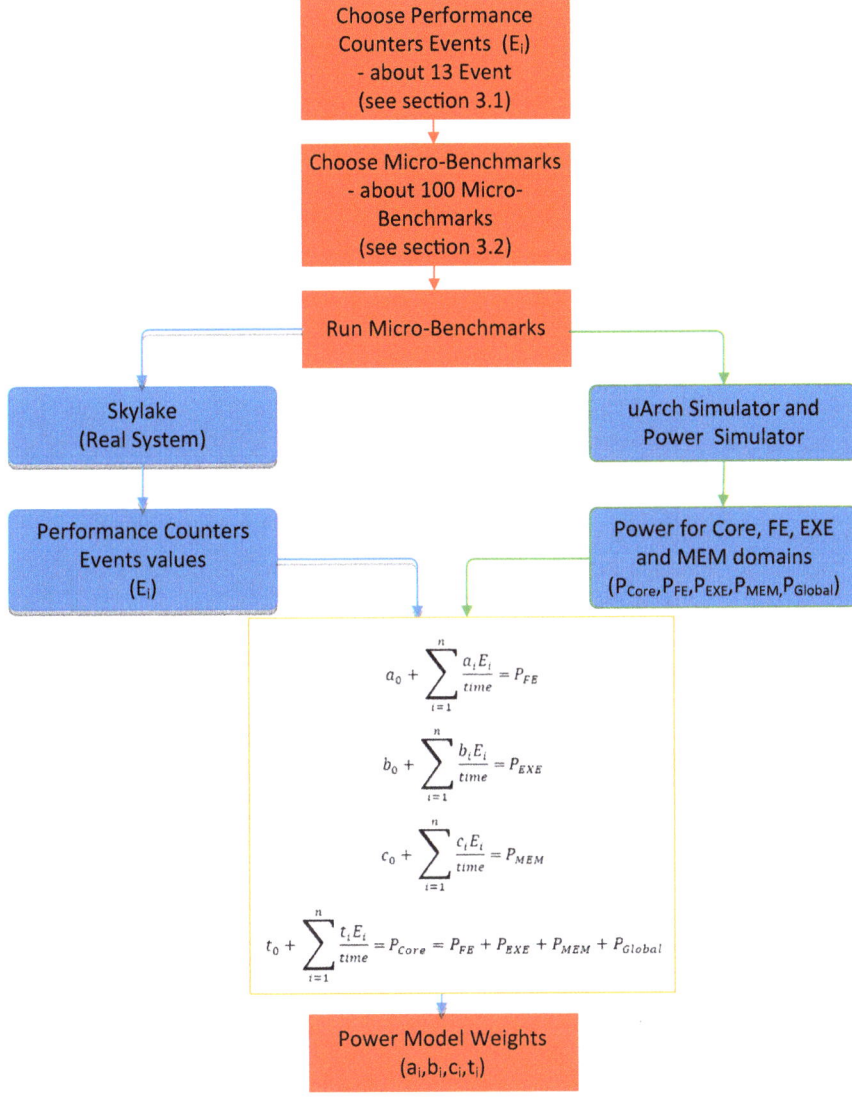

Fig. 3.2 Flow to determine the power-model weight based on simulator power breakdown and performance counters data

3.3.1   Events Selection

While prior works [17, 22, 23] adopted specific performance counters, we employed a process that started off an expert-driven selection of performance-monitoring events (counters) that was then empirically pruned using

Table 3.2 Performance counter used to estimate the core and domains power

Domain	Performance counter	Weight
Core	Intercept	541.7
	FP_ARITH_RETIRED.128B_PACKED_DOUBLE	−6.008
	FP_ARITH_RETIRED.256B_PACKED_DOUBLE	234.207
	FP_ARITH_RETIRED.256B_PACKED_SINGLE	68.73
	UOPS_DISPATCHED_PORT.PORT_1	286.368
	UOPS_DISPATCHED_PORT.PORT_2	325.855
	UOPS_DISPATCHED_PORT.PORT_6	150.967
	IDQ.DSB_UOPS	44.696
	IDQ.MITE_UOPS	164.858
	IDQ.MS_UOPS	218.486
	LSD.UOPS	−14.001
	UOPS_RETIRED.RETIRE_SLOTS	−26.606
	L2_RQSTS.MISS	1302.42
	UOPS_ISSUED.ANY	199.259
Frontend	Intercept	87.54
	UOPS_DISPATCHED_PORT.PORT_2	−84.597
	UOPS_DISPATCHED_PORT.PORT_6	107.628
	IDQ.DSB_UOPS	18.1747
	IDQ.MITE_UOPS	136.256
	IDQ.MS_UOPS	213.617
	LSD.UOPS	−77.274
	UOPS_RETIRED.RETIRE_SLOTS	−82.495
	UOPS_ISSUED.ANY	157.881
Execution	Intercept	203.014
	UOPS_DISPATCHED_PORT.PORT_1	158.128
	FP_ARITH_RETIRED.128B_PACKED_DOUBLE	63.0428
	FP_ARITH_RETIRED.256B_PACKED_DOUBLE	239.818
	FP_ARITH_RETIRED.256B_PACKED_SINGLE	84.2479
	UOPS_ISSUED.ANY	193.634
Memory	Intercept	127.209
	UOPS_DISPATCHED_PORT.PORT_2	263.475
	UOPS_DISPATCHED_PORT.PORT_6	78.8816
	L2_RQSTS.MISS	1426.51

an iterative process until the model average error is closer to our target average error of 5%. We also carried out sanity checks to make sure that the winning counters were reasonable per Skylake microarchitecture details. Table 3.2 shows the final list of performance counters that were chosen by our method.

Initial counters: A set of initial counters (23 counters) were chosen based on an expert knowledge of architecture and the available performance counters that the processor offers. These counters may be divided into three categories:

I. Counters at **inbound/outbound of a domain**; for example, the counter UOPS_DISPATCHED_PORT.PORT_2 counts all load/store operations received at inbound of memory domain while L2_RQSTS.MISS counts memory requests sent to off-core at same domain's outbound.

II. Coverage for **subunits inside a domain**; for example, the counters LSD. UOPS and IDQ.{DSB,MITE,MS}_UOPS represent the activity of the assorted fetch units in the frontend domain.

III. Counters **commonly used by prior work** such as cycle or Uops retired.

Obviously, the counters' space is limited to the list published by Intel in the Skylake Core Performance-Monitoring Unit (PMU), which entitles 157 raw counters in total [21]. Many of these do not apply to our study and were consciously de-selected. For example, 80 out of the 157 do not apply to our scope—e.g., deal with Transactional Memory (TSX [24]), or count off-core-related events like L3 miss, HW prefetches, large TLB pages, hyper-threads interaction, or aliases to other existent events. Another group of 22 counters deal with issues that are less significant for the SPEC CPU2006 benchmarks—e.g., TLB flushes or misses in all levels or instruction fetch misses related. We carefully selected the 23 counters out of the remaining 55 counters. For example, 13 counters provide a breakdown of L2 cache misses into per-request type; hence we manually selected an aggregated: L2_RQSTS.MISS. We did keep granular events that are likely to affect the inside core power consumption like in UOPS_DISPATCHED_PORT.PORT_*.

Supervised pruning: Solving the linear regression was an iterative process. At each step, we dropped one counter—the least to impact power estimation accuracy. We verified that dropping each counter was reasonable when relating to its functionality. For example, UOPS_DISPATCHED_PORT.PORT_0 counts ALU operations executed, so it made sense to be included in the execution domain and not in memory domain. Each of UOPS_DISPATCHED_PORT.PORT_{2,3} count load/store operations in a symmetric fashion. The linear regression indicated only one counter is significant for the memory domain (i.e., dependent variables). This can be verified as the two counters reported close values in all training micro-benchmarks.

Sanity checks: We verified the winning counters are proper for each domain:

In **Front end**—a counter was selected for each fetch unit as explained above. We can reason selection of UOPS_RETIRED and UOPS_ISSUED to quantify speculation impact since the front end domain is at the first stage of the pipeline making it most sensitive to speculation compared to other domains. UOPS_DISPATCHED_PORT.PORT_6 counts branches which also impact front end behavior, for example, lookup in branch prediction structures.

In **Memory**—UOPS_DISPATCHED_PORT.PORT_2 and L2_RQSTS.MISS at inbound and outbound of the domain were selected as expected.

UOPS_DISPATCHED_PORT.PORT_6 was included since it well correlates with high-IPC (probably as a proxy for memory accesses hitting in the L1 data cache).

In **Execution**—ports 0 and 1 generally accommodate similar execution units such as integer ALU and FP arithmetic [9]. UOPS_DISPATCHED_PORT. PORT_1 was selected in the process; PORT_0 was very close hence dropped. The FP_ARITH_* counters provide scalar versus vector (width) versus precision distinction also made sense. Note also that the wider the vector size, or the higher the FP precision, the bigger the cost assigned per operation.

3.3.2 Training-Set Selection

A set of over hundred kernels were carefully selected for the linear regression. The designated collection of kernels has good coverage for both the architecture and the microarchitecture. It includes integer, floating-point (single and double precision), vector (SSE, AVX, and AVX2 [25]), and FMA instructions, crossed with various microarchitecture parameters such as cache miss rate. We also added kernels that exercise specific units of the microarchitecture such as the assorted fetch units.

The process of selecting the micro-benchmarks for the training set is similar to the events selection process that we described in the previous subsection. The process started off an expert-driven selection of micro-benchmarks that was then empirically pruned using an iterative process. The micro-benchmarks were written in order to cover both the architecture and the microarchitecture feature and domains of the Skylake processor. The coverage of the features was monitored using the events that were selected previously. If some event was zero then we wrote dedicated micro-benchmarks to cover it, for example, the LSD.UOPS counters were zeros at the first set of the training micro-benchmarks, we wrote a dedicated micro-benchmark with a short loop that inters inside the LSD micro-operations cache and covered this event.

3.3.3 Reference and Estimated Power Calculation

We used an Intel proprietary simulator to obtain the reference power numbers for the core and its sub-domains. The simulator is built using microarchitecture events, these are obtained from a Microarchitecture Simulator (uArch-Sim) that runs before the power simulator; the power simulator uses the microarchitectural events and low-level design data (such as capacitance, or design block activity factor). The simulator uses power estimates from previous CPUs and circuit simulations on micro-benchmarks to estimate the power of architectural events (e.g., register access, cache access, ALU operation) and FUB level. The power per each event is multiplied by the activity of the block in order to estimate the power. Core sub-domain, such as FE, has many FUBs so to estimate the power of the frontend (for example) the power simulator uses hundreds of events. Summing all frontend sub-domain FUBs power will give the frontend total power. The accuracy of the Intel power proprietary simulator is 3–5% for performance workloads (i.e., workloads without CPU idleness).

For each counter, we assigned a weight in the equation to calculate the core domain's power. Equation 3.2 shows the linear function used, PerfCounter$_i$ is the value of the performance counter normalized by the sampling interval, the sampling interval is obtained from the ref_cycles fixed performance counter, dividing the ref_cycles by the reference-clock frequency (fixed clock and not part of the DVFS domain) gives the actual runtime. Each normalized performance counter value is multiplied by the weight$_i$ corresponding to the counter.

3.4 Power Estimation Tool

Our tool estimates the power of the core and breaks it into three sub-domains of the core: frontend (FE), execution (EXE), and memory (MEM). The counters used to calculate the power of each one of the components are different: these counters were determined according to the process described in the previous section and they showed the closest correlation to the components' power. The counters for each component are listed in Table 3.2. Note that some counters have negative weight, this is due to overlapping (e.g., shared resources) in sub-domains that different counters target.

To collect the performance counter, we used Linux perf [26]; this handles the configuration of the performance counter and can report its value. Besides, it features counter-event multiplexing-simplified profiling when the number of performance events is higher than the physical counters supported at the processor.

We used a set of representative micro-benchmarks that stress and train each part of the core domains. We provide some details in Sect. 3.3.1.

3.4.1 Tool Auto-calibration

The Intel simulator does not supply the absolute power number for the running micro-benchmark, it rather gives an output that represents the dynamic capacitance (αC_L in Eq. 3.1, also denoted C_{dyn}) that each domain showed while running the micro-benchmark. To convert that value to real power we need to know other parameters like voltage and frequency. The frequency is available to us but the voltage level is not visible to the user and it varies from one processor package to another. To extract the calibration values, we used the RAPL output which is already calibrated at the manufacturing process to report the correct power value. We did the calibration using a micro-benchmark that does not have L2 misses to make the RAPL reporting more accurate. The calibration is carried out by running the micro-benchmark on the simulator which reports C_{dyn}. We then ran the micro-benchmark on the real Skylake system where the micro-benchmark runs many times in a loop in order to have higher accuracy of the RAPL. The RAPL value is read and divided by runtime to estimate what is the actual average power that was consumed by the micro-benchmark. We then divide that average power by the C_{dyn} to get the *scaling factor* that will be used to convert C_{dyn} inside our tool into actual power.

The *scaling factor* is machine specific as it depends on the voltage–frequency curve of the specific processor. Our tool carries out the calibration process automatically and estimates the *scaling factor* at the tool initialization stage. This provides a significant advantage where our tool can auto-run on any product proliferation (aka SKU) based on the Skylake microarchitecture, not necessarily the specific system we experimented with. For example, Desktop or client models, Core i7 or Core m5, any frequency, etc.

3.4.2 Model Accuracy Validation

While we have high confidence in our breakdown model; it is not easy to find a reference model for the power of fine-grain components on real systems. To compare our tool with larger programs such as SPEC, we cannot use the power simulator. Instead, this subsection recaps on various methods we used to verify the model accuracy.

Validating the model accuracy requires reference power model for the core and its sub-domains. We have used two reference models for this purpose, the first uses the Intel proprietary power simulator for the breakdown validation, while running on short programs (kernels), and the second reference model is the RAPL [13] energy reporting register for the core energy (PP0_ENERGY_STAUS MSR), while running on larger programs such as SPEC.

3.4.2.1 Accuracy Versus Power Simulator

We have validated our model and tool for power breakdown against the power simulator, the process of the validation is described in Fig. 3.3. A set of micro-benchmarks (listed in Table 3.3) were used, each benchmark ran on the power simulator which outputs two sets of data:

(1) **Trace with microarchitectural events**—The Intel proprietary power simulator uses the microarchitectural events and low-level design data (such as capacitance, design block activity factor, etc.) in order to estimate the power at sub-FUB levels.
(2) **Performance Counters**—The counter values are consumed by our power model in order to estimate the power per each domain.

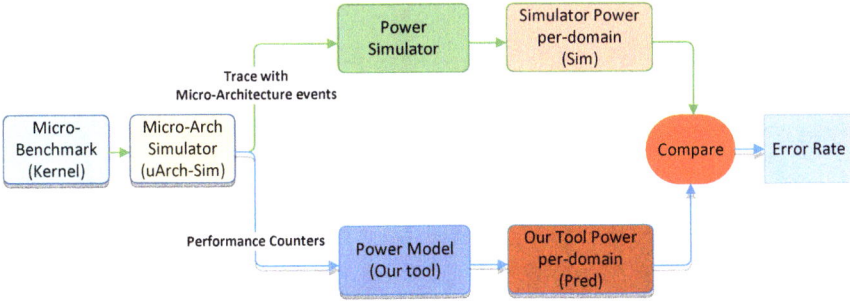

Fig. 3.3 Power-model validation process

Table 3.3 Micro-benchmarks (Kernels) used for validating the power model

Kernel name	Kernel description
DAXPY	Double-precision a * X + Y kernel. Commonly used in matrix multiplication
Mulss	Multiply scalar single-precision floating-point instruction in a loop
Spline_int_128bit	Spline interpolations (128bit)
Divpd	Packed double-precision floating-point divide instruction in a loop stressing latency aspect
DGEMM_MKL_256bit	Multiplying matrices using the dgemm routine at Intel Math Kernel Library (MKL), which calculates the product of double-precision matrices
Sqrtpd	Compute square roots of packed double-precision floating-point instruction stressing latency aspect
UintToDouble_256	Converts vector unsigned integer elements into double using vector instructions (256 bits)
Transpose_128bit	Matrix transpose using 128 bits vectors
Add_Two_vec_256bit	Add two vector instructions of 256 bits

The results are shown in Fig. 3.4, both the power simulator (Sim) and the predicted power by our tool (Pred) are shown. In addition, the figure includes the average and maximum error across all domains for each micro-benchmark. We can see that the error rate is up to 7% compared to the Intel proprietary power simulator on the micro-benchmarks comparison. The error is due to multiple sources; this includes the linear regression inherited error. In addition, the data switching factor can lead to variations in the actual power consumption, hence it is a source of error in our model.

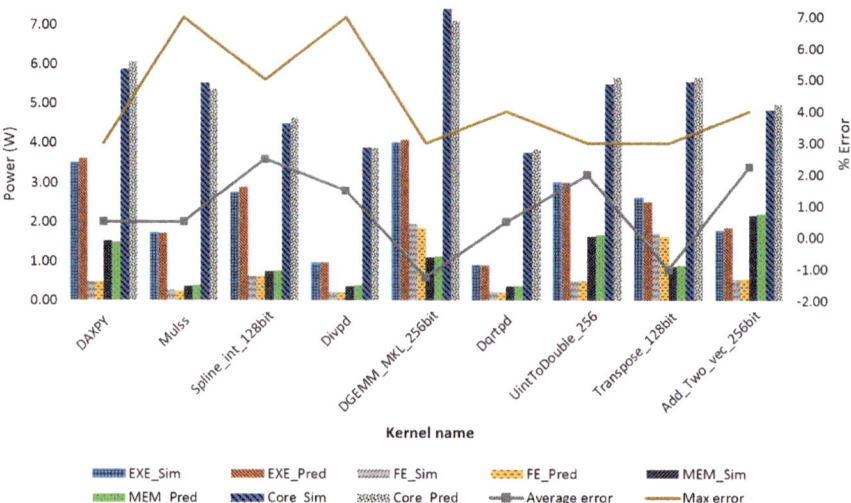

Fig. 3.4 Power per-domain outputted by power simulator (_Sim) and Our tool (_Pred) on various micro-benchmarks

3.4.2.2   Accuracy Versus RAPL

To compare our tool with larger programs such as SPEC, we cannot use the simulator, because running large programs on the simulator will require days of simulation time and very large storage space. For this purpose, we have validated the tool against RAPL [13] energy reporting register for the cores' energy (PP0_ENERGY_STAUS MSR) on a real system. This MSR reports the energy consumed by the different cores and the LLC—it is the closest available to the granular levels our work models. Our tool does not attempt to model the LLC power (out of the core domain). For validating the power consumption using RAPL, we looked at benchmarks that do not generate L2 cache misses and that work on single core. Results showed an error rate between 7 and 20% for benchmarks with near-zero L2 cache misses. It is important to note that using RAPL to measure the power of a single core is not fully accurate, first because the error rate of RAPL itself is up to 5% [13]; second, RAPL reading in our case estimates the power of all cores and LLC, if the cache is idle (no L2 cache misses) does not mean that its power is zero as the cache has static power and is not fully power gated. Moreover, additional noise can arrive at RAPL reading from operating-system threads. If some OS thread wakes up, it can change the RAPL readings. Despite the RAPL inaccuracy, we wanted to show that our model correlates well with the RAPL reporting.

Figure 3.5 plots the core power from our model, RAPL PP0 power, alongside their ratio in bars for all SPEC CPU2006 INT and FP benchmarks sorted by the ratio. Additionally, we included the L2 cache miss rate and its trend (dashed line). We measure L2 cache misses of any request type. The bars use the left-hand Y axis,

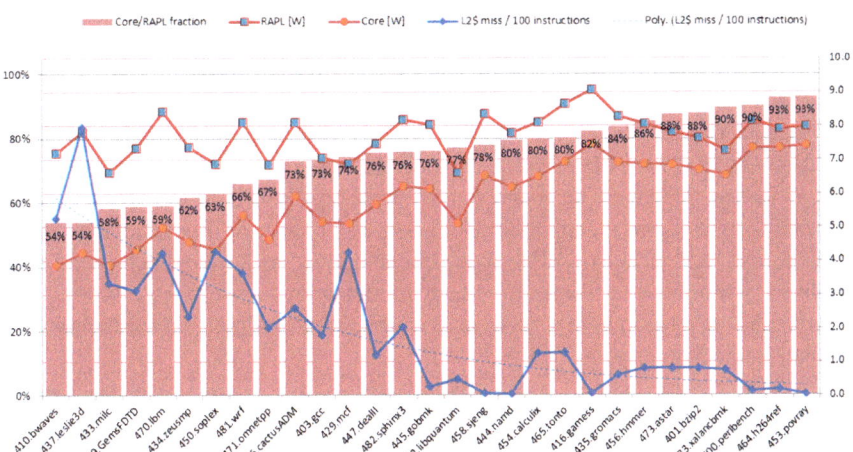

Fig. 3.5 RAPL power versus core estimated power

while all others use the right-hand Y axis. The graph suggests that the lower the L2 cache misses the closer is our core estimated power to the RAPL "reference" power. The correlation between the two variables is 68% for all benchmarks and 75% for those with L2 cache miss rate below 0.5.

One source of inaccuracy we noticed for a subset of SPEC FP benchmarks: gromacs, games, namd, GemsFDTD, milc. These benchmarks have significant use of FP scalar (nonvector) operations—over 20% of retired uops are FP scalar compute operations—and so are likely to have their execution domain and core power underestimated.

3.4.2.3 Performance Profile Cross-check

Lastly, we closely examined and cross-checked the power breakdown against performance breakdown (using the TMAM-validated method [20]) in addition to designated performance counters where applicable. This is extensively illustrated in the Sect. 3.5, for which a sizable part of this work was devoted.

3.4.3 Over-Time Estimation

Our tool can be configured using a command line option to report power estimation over-time, the tool can report the average power consumed every given time interval (e.g., 100 ms, 1 s …). This option gives better understanding and visibility to the user about the behavior of the examined benchmark at different phases. A sample output is provided in Sect. 3.5.6.

3.4.4   The Scope of the Tool

The scope of this work is on single-threaded CPU-bounded workloads, i.e., performance workloads that do not include CPU idleness (C-states [9]). Moreover, our tool assumes that the CPU performance state (P-states [9]) and temperature do not change in the middle of the benchmarks runs, i.e., DVFS in the middle of the benchmark run is not supported. In addition, our tool uses auto-calibration methodology that enables porting it to different processors, frequencies, and temperatures. Currently, a single frequency/temperature is supported during the benchmark run) without the need to recalibrate manually if, for example, the voltage–frequency curve is changed.

The tool can be enhanced in the future to include multiple frequencies and temperatures by applying a calibration process at the initialization phase to multiple frequencies/temperatures and keep a mapping of scaling factors per each frequency or temperature. This enhancement requires inquiring the processor frequency/temperature periodically and using the scaling factor corresponding to the current frequency/temperature point.

The tool currently supports a single core and it can be extended to multi-core, as the performance events reading today is per-core. Few changes are required during the calibration process and tool structure to support multi-core power breakdown at sub-core domains.

3.5   Experiments and Results

In this section, we present experimental results of the method on a Skylake system. First, we demonstrate how some compiler optimizations affect the breakdown on simple kernels. Then, we present characterization data for the SPEC CPU2006 benchmarks on our system with the setup described in Table 3.3. We carry out a global analysis of the data and a drill down into a few case studies. Lastly, we share overall results on model accuracy.

3.5.1   System Setup Parameters

For the experiments in this study, we used a Skylake platform with the following setup. We also used pmu-tools/toplev [27] to obtain the full TMAM performance breakdown used in the case studies section.

Temperature: Preliminary experiments have shown that a rise in processor temperature can increase power by as much as 0.5–0.8 W. To minimize this impact, benchmarks are run continuously for a 5-min warmup period to allow the temperature to stabilize before starting the experiment. We do not expect to have much

Table 3.4 Baseline system setup parameters

Processor	Intel® Core™ i7-6700 K. 3 GHz fixed frequency, 8 MB L3 cache. Hardware prefetchers disabled[a]
Memory	8 GB DDR4 @ 2133 MHz
OS	Ubuntu 14.04, Linux kernel 3.19, Hugepages disabled perf version 3.19.8-ckt5
Benchmark	SPEC CPU2006 v1.2 (base/rate mode)
Compiler	Intel Compiler 14.AVX2 ISA with FMA. INT/FP target 32/64-bit modes, resp

[a]Due to erratum SKL095 [33] on OFFCORE_RESPONSE the visibility to hardware prefetches is too limited to enable proper power modeling

variation in the temperature after the warmup time as the workloads assumed to be CPU bounded and no idleness expected (Tables 3.4, 3.5, and 3.6).

3.5.2   Sample Optimizations

We start with very simple kernels in order to reason our power modeling and see how common compiler optimizations affect power and performance.

Kernels (or micro-benchmarks) are tight loops of specific instructions. We choose different instruction types for implications on execution domain power. Similarly, loop unrolling is used to check implications on the frontend and memory domains. Lastly, we examine the Fused Multiply–Add (FMA) instruction which has been recently introduced by Intel's AVX2 and its efficiency in the Skylake implementation.

3.5.2.1   Varying Code Size

In this experiment, we built a micro-benchmark with simple loop of MOV and ADD register-to-register operation (no memory access). Then we applied loop unrolling with factors of 1,2,4…2048 as shown in Fig. 3.6.

Figure 3.6 through Fig. 3.8 report relative performance (time), power, and energy for the various unrolling factors in the X axis. The bars are the per-domain power breakdown, and the total bar height represents the core power relative to baseline (leftmost bar). The same annotation is used in figures later in this subsection.

The execution power, relative to baseline, is fairly constant as the unrolling factor increases. On the other hand, the frontend power increases with a bigger unroll factor. The reason is that different fetch units (as depicted in Fig. 3.1.) deliver the instructions as the code size increases. For example, small loops can fit nicely in the Loop Stream Detector (LSD), which is power efficient as most of the other

Table 3.5 Comparison of recent power estimation methods

Work	Scope		Accuracy	Granularity	Reporting
Bertran [18]	System	Intel Core 2 Duo	~6%	Calibrated using SMAPI[ThinkPad]	Medium–High: Dedicated per group of microarchitectural components	Power in watts
CAMP Powell [17]	Simulation	(Intel Core 2009)	10–15%	Calibrated against ALPS propriety model	High: Dedicated per-unit	Pseudo-power [Pseudo-watts]
ISCI [22]	System	(Pentium 4, 2003)	~13%		High: Dedicated per-unit	Power in watts
Spiliopoulos [34]	System	(Nehalem 2008)	5–15.1% (average 3.9%)		Very Low (No breakdown): whole CPU power	Power in watts
RAPL [13]	System	(Sandy Bridge 2011)	5%		Low (Coarse): Cores + LLC, Graphics, platform, DDR	
Ours	System	(Intel Core, Skylake 2015)	5–10%	Calibration with RAPL	Medium: Core sub-domain (group of units)	Power in watts

Table 3.6 SPEC CPU2006 power and performance characterization data on Skylake

Benchmark	Power [W]					Performance	
	RAPL/PP0	Core	Front-end	Execution	Memory	IPC	FLOP/cycle
400.perlbench	8.2	7.4	1.6	3.7	1.5	2.76	
401.bzip2	7.6	6.7	1.2	3.1	1.5	1.73	
403.gcc	7.0	5.2	1.1	2.3	1.2	1.29	
410.bwaves	7.2	3.9	0.6	1.8	1.0	0.58	0.47
416.gamess	9.1	7.5	1.6	3.9	1.6	2.66	0.79
429.mcf	6.9	5.1	0.9	2.0	1.3	0.78	
433.milc	6.6	3.9	0.6	1.7	1.0	0.65	0.44
434.zeusmp	7.4	4.6	0.7	2.3	1.0	0.86	1.42
435.gromacs	8.3	6.9	1.2	3.4	1.4	1.66	2.24
436.cactusADM	8.1	5.9	1.0	3.1	1.3	1.18	2.75
437.leslie3d	7.8	4.2	0.6	1.9	1.2	0.58	0.86
444.namd	7.8	6.2	0.9	3.4	1.1	1.88	1.03
445.gobmk	8.0	6.1	1.7	2.6	1.2	1.25	
447.dealII	7.5	5.7	1.0	2.8	1.3	1.74	0.25
450.soplex	6.9	4.3	0.7	1.8	1.1	0.62	0.07
453.povray	8.0	7.4	1.8	3.6	1.5	2.37	0.52
454.calculix	8.1	6.5	1.1	3.5	1.3	1.94	1.63
456.hmmer	8.1	6.9	0.9	3.6	1.7	2.51	
458.sjeng	8.4	6.5	2.0	2.9	1.2	1.75	
459.GemsFDTD	7.4	4.3	0.6	2.1	1.1	0.84	0.56
462.libquantum	6.6	5.1	0.4	3.0	1.0	1.86	
464.h264ref	7.9	7.3	1.5	3.6	1.7	2.56	
465.tonto	8.7	7.0	1.4	3.6	1.5	2.39	0.92
470.lbm	8.4	5.0	0.8	2.6	1.1	0.83	1.57
471.omnetpp	6.9	4.6	1.0	1.9	1.1	0.84	
473.astar	7.8	6.9	1.3	3.0	1.5	1.48	
481.wrf	8.1	5.4	0.9	2.6	1.3	1.14	2.09
482.sphinx3	8.2	6.2	1.0	3.3	1.3	1.83	0.94
483.xalancbmk	7.3	6.5	1.5	3.1	1.5	2.40	

frontend blocks are switched off [9]. This explains the small frontend power in the two leftmost bars. As the code size increases, it gets delivered by less efficient fetch units, power, and performance-wise. At the unroll factor of 512 is where the code misses the L1 Instruction cache (L1I) and fetches instruction lines from L2 cache. There is a noticeable power of the Memory domain which accounts for power consumption of the L2 cache accesses.

Overall, we can see a higher unroll factor (which resembles a bigger code size) drawing higher power while the performance also decreases. In this case, managing

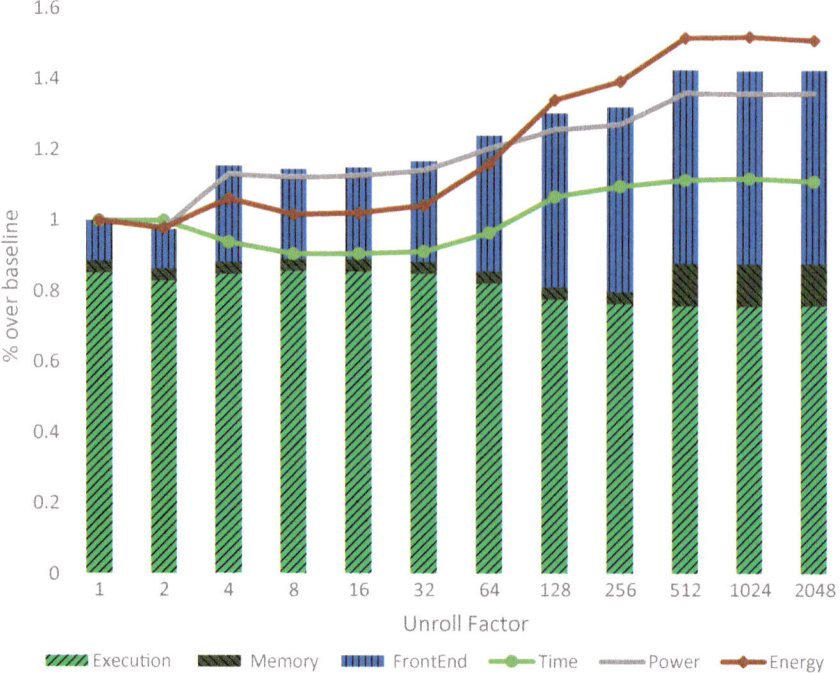

Fig. 3.6 Power breakdown and performance of unrolled Scalar ADD kernel

the unroll factor (or reducing code size) will give us optimization for power and performance together. That is, this example demonstrates that optimization for performance can be also an optimization for power.

3.5.2.2 Vectorization Impact

In this example, we ran the same simple ADD-MOV micro-benchmark on a scalar, SSE (128 bits), and AVX (256 bits) Instruction Set Architectures (ISA) [9].

In Fig. 3.7, we can see that the performance of SSE is better than the scalar case while the AVX has the best performance. SSE improves performance by almost 4x since the kernel uses single-precision floating-point accuracy (32 bits). The power of AVX is 42–65% higher than baseline (scalar) as expected since the vector operations consume more power (multiple operations versus single operation in scalar). This is explained by the bigger execution bars as we move from scalar to vector 128-bit to vector 256-bit categories. The code size (unrolling factor) has a similar impact as we observed *in* Fig. 3.6 which suggests the frontend power works on more instructions set.

Note that overall energy with AVX is better than the baseline. This demonstrates that while vector instructions can consume higher power, they are more

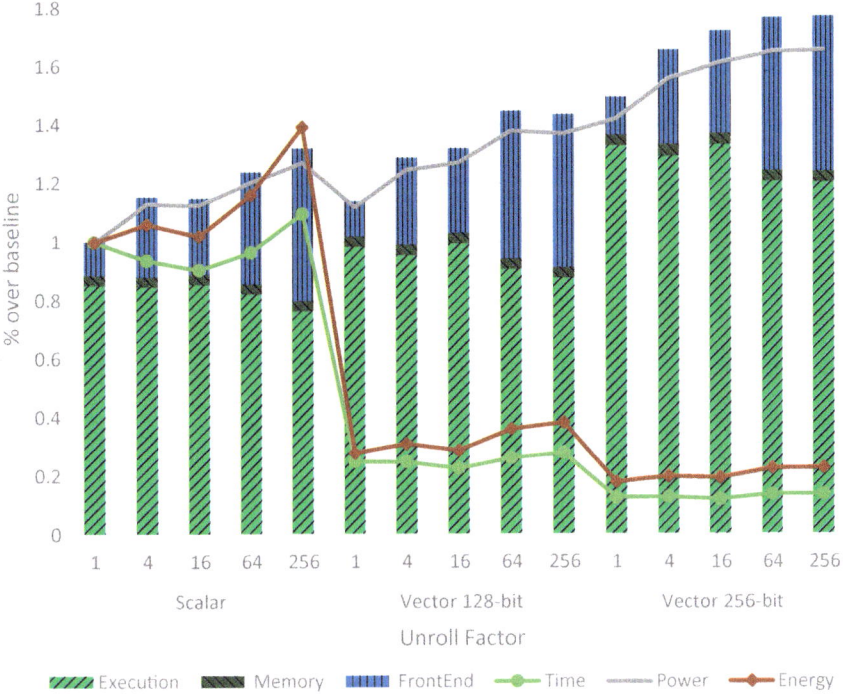

Fig. 3.7 Scalar, vector 128- and 256-bit versions of ADD kernel at different unrolling factors

energy-efficient than nonvector scalar instructions. That is, vectorization is an energy-efficient optimization on the Skylake microarchitecture.

3.5.2.3 The FMA New Instruction

In this experiment, we took two micro-benchmarks with a simple loop that does multiply–add between registers (1 mov, 2 operations, no memory accesses). The first micro-benchmark does the multiply–add using two distinct instructions, while the second one uses the recently introduced FMA instruction [24]. Both kernels do the same amount of work.

In Fig. 3.8, we can see that the performance of the FMA is better by a factor of 2x than multiply–add using distinct instructions. The overall power has increased by $\sim 10\%$ for the middle unroll factors. Hence, FMA has better net energy than a none-fused version.

When we look at a power breakdown, the execution power is slightly better in the FMA case (or another way of looking at it is that the power of the distinct multiply–add is higher), possibly since fewer net operations are sent to the

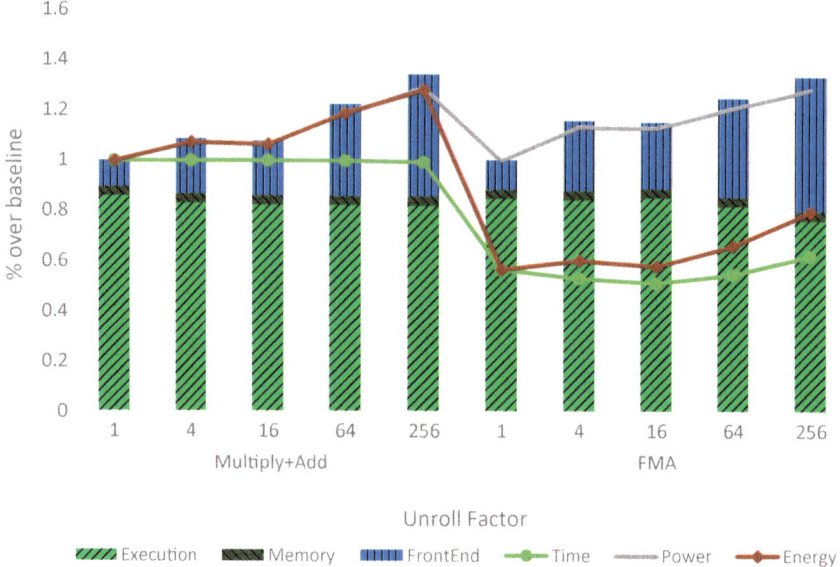

Fig. 3.8 Scalar multiply–add and FMA kernels at different unroll factors

execution units. The code size (unrolling factor) has a similar impact like we observed in Fig. 3.6.

3.5.3 Floating-Point Apps

We ran our model on SPEC CPU2006 FP benchmarks built with an Intel Compiler. The next two figures are twofold: the upper part represents a power profile with absolute power (solid lines in Watts) for the package and core as well as the breakdown inside the core (bars are fractions). The lower part is the performance profile with the absolute performance—Instructions Per Cycle (IPC)—and its breakdown using TMAM [20] with the representation is described in Sect. 3.2.5. The benchmarks are sorted by increasing retiring from left to right. Retiring is the fraction of pipeline slots utilized by operations that eventually retire (i.e., from the program's true path). Generally, the higher the retiring bucket, the higher IPC.

Notice the RAPL [in watts] is retrieved based on RAPL PP0 counter (see Sect. 3.2.4) provided by the processor hardware (that is not relying on our method), while core is modeled by our tool.

In Fig. 3.9, the power lines in the upper graph correlate with the performance whenever the core power is the greater part of the package power (RAPL). That is, *the general trend is that core power correlates with IPC*. This makes sense as one would expect the higher the throughput of operations executed by a processor

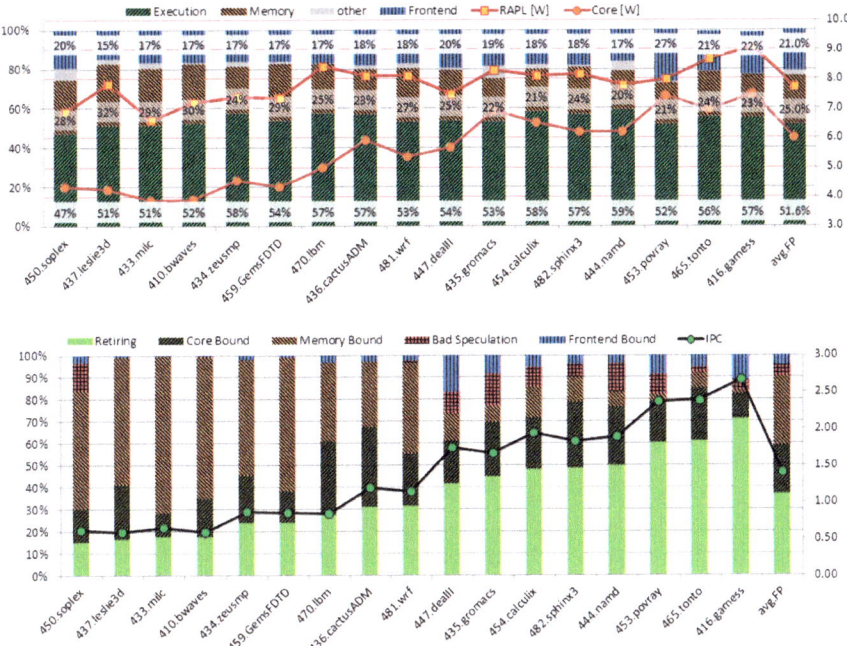

Fig. 3.9 Power (upper) and performance (lower) profiles of SPEC CPU2006 FP apps

(IPC), the bigger the power consumed to do that. One exception to this trend is evident with the move from 447.dealII to 435.gromacs where the power increases while performance decreases. An inverse of this phenomenon occurs for 433.milc though to a lesser extent. We study both the cases in Sects. 3.5.5.1 and 3.5.5.2.

On average, the core is the biggest power consumer within the package for these FP workloads where it consumes 71% of the power.

One interesting insight is that the frontend consumes 21% of core power, while it is an insignificant performance bottleneck—frontend bound is 5% on average. This can be explained as the frontend portion of the pipeline must always fetch instructions to feed the backend, regardless whether the fetch operation itself is critical performance-wise or not. Since compiler's code generation heavily impacts the frontend footprint, hopefully, this means there is a room for energy efficiency. Profile-guided optimizations may be promising in this direction if they can reduce the power consumption of the frontend without degrading performance. The LSD is one example hardware optimization as demonstrated in Sect. 3.5.5.3.

For benchmarks on the left side (whose IPC is below one), our model suggests core power is not the significant part of package power. This corroborates with the performance profile where these benchmarks get low IPC since they are significantly memory bound. Examining toplev [27] results for next levels in the TMAM hierarchy, suggests the performance bottleneck is in areas outside the core—LLC and external memory (we omitted these results due to space limits). Notice the lower core power fraction suggests the design is somewhat efficient where the cores do not burn much power while waiting for the off-core memory subsystem. In particular, we see the memory power fraction for them is in the 24–32% range of core power, which is not too far from the average (25%).

3.5.4 Integer Apps

The INT benchmarks show a similar *general trend where the core power correlates with IPC* in Fig. 3.10. See the previous subsection for graph annotation. A key exception case is 462.libquantum which has a noticeable drop in power compared to its predecessor benchmark and yet has better performance—we study this case in Sect. 3.5.5.3.

On average, the core consumes 82% of the package power for the INT benchmarks (recall FP was 71%). This can be explained by examining the TMAM performance profile. INT benchmarks are less bounded on memory performance-wise (average memory bound is 15% for INT vs. 31% for FP). This means more of the power is likely to be consumed by components inside the core, akin to what was discussed for memory-bound FP benchmarks in the previous section. Moreover, INT benchmarks have a higher fraction of bad speculation (15%

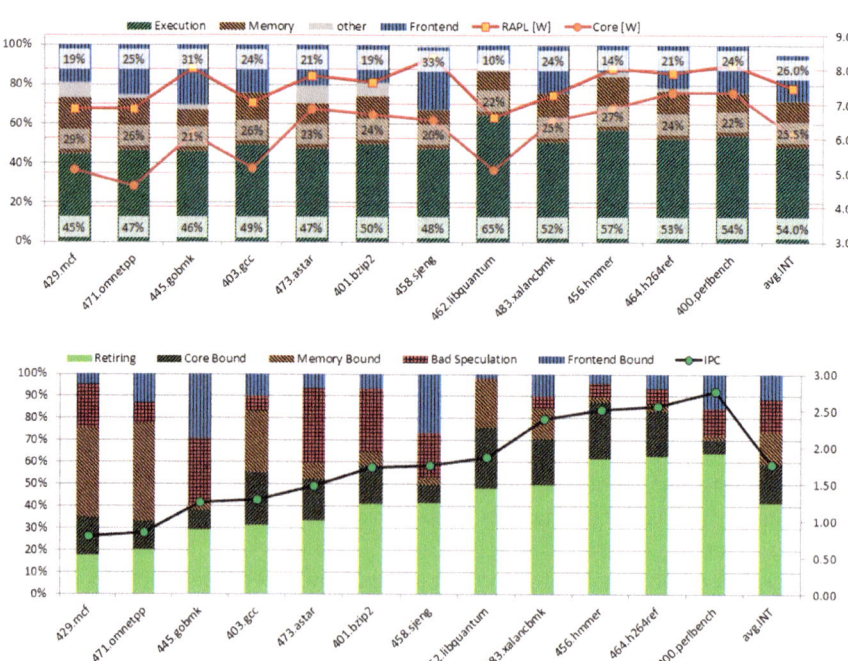

Fig. 3.10 Power (upper) and performance (lower) profiles of SPEC CPU2006 INT apps

vs. 5%), which in turn means the core requires more power while executing operations out of an incorrect program path or when recovering back to the true path afterward.

Looking at the power breakdown, INT benchmarks show higher execution domain power fraction compared to their FP counterparts. This is most likely due to the higher average IPC (1.76 vs. 1.4). Also the frontend power fraction is higher, since INT benchmarks are known to be more sensitive performance-wise on instruction fetch and branch predictions [20].

3.5.5   Case Studies

In this section, we examine some interesting cases of the SPEC CPU2006 results.

3.5.5.1   435.Gromacs

435.gromacs consumes core power of 6.9 W as reported by the data of Table 3.6 in the Appendix. Figure 3.9 suggests this is a significant increase over its predecessor (447.dealII with 5.7W core power). On the other hand, the IPC has slightly decreased when comparing the same two benchmarks.

We use the Floating-Point Operations per Cycle (FLOP/C) metric to explain this. One can claim it is a better metric to represent FP performance than plain IPC, since it properly accounts for vectorization throughput. For example, a 256-bit vector ADD instruction can serve up to 8 single-precision FP elements. It is counted as one in IPC versus eight in FLOP/C.

Table 3.6 reports the FLOP/C and IPC metrics for all FP benchmarks. Note FLOP/C is at 2.24 for 435.gromacs. This means while 435.gromacs executes instructions at a slightly lower rate compared to 447.dealII, it delivers FLOP/C at a much higher rate (by a factor over nine). Our execution domain power model correctly reflects this case: it has the higher absolute power of 3.4 W in 435.gromacs versus 2.8 W in 447.dealII (20% higher). Microarchitecture wise, the execution cluster is where the power consumption manifests when executing vectors versus nonvector operations.

3.5.5.2   433.Milc

An inverse of the 435.gromacs phenomenon occurs for 433.milc: it has higher performance and lower power compared to its predecessor (437.leslie3d) in Fig. 3.9. We use FLOP/C once more to explain this with data from Table 3.6 in the Appendix.

While 433.milc has higher IPC, it actually has a lower FLOP/C metric (by a factor of almost 2) compared to 437.leslie3d. Examining the performance counters of the Skylake processor, we observe that only 8% of the arithmetic instructions of 433.milc has actually used vector ISA, compared to 95% for 437.leslie3d. The power model indicates less execution and core absolute power for 433.milc over 437.leslie3d (see Table 3.6). Again, our model correctly reflects this case, despite the fact that the core power is not the biggest consumer of package power for these two benchmarks.

3.5.5.3   462.Libquantum

Recall that the general trend was that core power correlates with IPC in Fig. 3.10. 462.libquantum is a positive outlier where its IPC is slightly better than its pre-decessor benchmark (458.sjeng), while it has quite good core power reduction (6.5 W for 458.sjeng down to 5.1 W for 462.libquantum per Table 3.6).

The power breakdown clearly shows the frontend power was reduced from 33% (2.0 W) down to 10% (0.4 W). This has occurred since almost all of the instructions of 462.libquantum were actually delivered by the LSD unit which is the most efficient fetch unit in frontend as indicated in Sect. 3.5.2.1. In comparison, only 1% of 458.sjeng instructions were delivered by the LSD.

3.5.6   Trace Over Time

Our tool can be configured using the command line option to report power esti-mation over-time, the tool can report the average power consumed for every given time interval (e.g., 100 ms, 1 s …). This option gives a better understanding and visibility to the user about the behavior of the examined benchmark in different phases.

Figure 3.11 shows trace over time for the 416.gamess benchmark as outputted by our tool. The graph also includes the sampling over time of the Decoded Stream Buffer (DSB) coverage performance metric. The DSB serves as decoded uop-cache that allows the frontend to be run at lower power in case the uops stream out of the DSB. The figure shows that the frontend power is relatively high at the interval of 0–110 s, while the power reduces at the middle interval that lasts till second 470. We can observe that the power of the frontend is high when the DSB coverage is low and vice versa.

3.6   Related Works

Frank Bellosa's work [16] shows the linear correlation of hardware events and energy. Then, they use hardware activity to establish a thread-specific energy consumption model. Since then, a lot of work has been done for power estimation

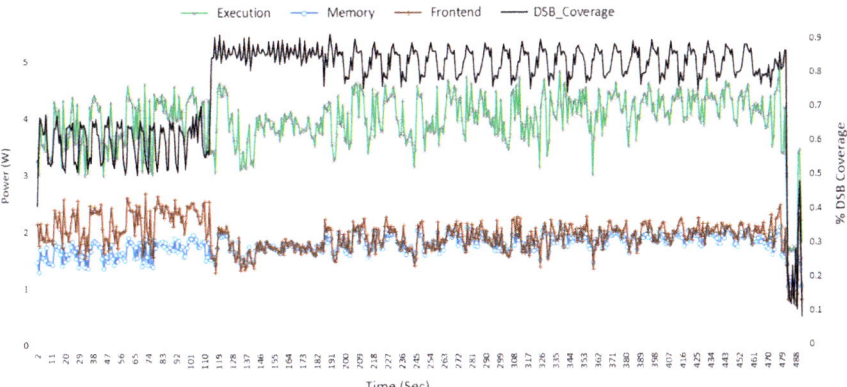

Fig. 3.11 Trace over time for the 416.gamess with 1000 ms interval

based on hardware performance counters. PMCs-based power estimation models in earlier work can be divided into two categories [19]. The top-down method uses a reduced set of events (usually around 5), aiming to build a fast and simple model with small overheads, while the bottom-up approach breaks down the power components based on microarchitecture (it can use hundreds of microarchitectural events). The latter produces a more accurate power model by gathering more information to reflect the power characteristics of applications, at the cost of increasing the complexity of modeling. Our model balances between accuracy and cost, and can be viewed as a moderation on both sides (we used about 13 events).

Bertran et al. [18] present a decomposable power model for Intel Core 2 Duo processor. They use a total of 13 counters for estimating the power consumption of different microarchitectural components. They use micro-benchmarks and incremental linear regression technique to come up with individual power consumption figures for components such as frontend, Integer execution units, FP units, BPU, and SIMD units. A later work by Bertran et al. [19] reduces the problem of generating decomposable power models for different architectures by proposing a systematic modeling methodology. The method used by [18] for power measurement using SMAPI [28] is platform specific: it uses the battery charging information to estimate power. This method cannot be applied to desktop and is less accurate relative to RAPL which measures the power for finger domains and not for the whole platform. In addition, the method at [18] has a restriction that it cannot map microarchitectural components to power components if the activity of the microarchitectural components cannot be isolated using dedicated micro-benchmarks, in our method, we use the power simulator data that have accurate power estimation per component. Moreover [18] does not have an auto-calibration capability; this means that when moving to another processor (same architecture) or applying a different frequency or temperature the user needs to regenerate the model for the new setup.

Powell et al. [17] proposed a model that uses 9 PMCs to estimate the power of various core sub-domains of Intel-like core architecture; they used a power simulator and an architecture simulator to produce the model which was built using linear regression on the data obtained from the power and architecture simulators. The model reports pseudo-watts and not actual power. However, comparing with our tool, we have implemented the tool on real systems and our tool reports read power (in Watts). In addition, our tool does auto-calibration that enables porting it to another processor or running at different frequencies. Moreover, the model uses 9 PMCs to estimate the power of more than hundred core subunits; this is less accurate than our model that uses a dedicated counter for each subunit that best represents it. The pseudo-power reported by the model from [17] is compared to a simulator which is itself 5–10% accurate. Our tool runs on a modern real processor and was compared against the RAPL reporting.

Karan Singh et al. [23] achieves a runtime per-core power estimation of multithread and multi-program workloads using the top-down method [29]. The model estimates the power for each core at the thread level, not at sub-core domains. They categorize the processor's hardware event into four classes (because their environment platform has only four performance counters). Then the topmost one is chosen in each class which is the most correlated to power. With the runtime data from executing micro-benchmark, they built a piece-wise linear mode and achieved median errors of 3.9%, 5.8%, and 7.2% for the SPEC-OMP, NAS, and SPEC CPU2006 benchmark suites respectively. However, compared with ours, their model is estimating at lower granularity (thread level) while we model the per-core domains. Furthermore, their model uses relatively few counters (4 counters) while we use dedicated counters for each sub-domain of the core; we used 13,8,5,3 counters to model the power of the whole core, front-end cluster, execution cluster, and memory cluster respectively.

Isci and Martonosi [22] decompose CPU into 22 power breakdowns based on a function unit which is a typical bottom-up approach [29]. They estimated the maximum power for each of the units according to physical area based on Pentium 4 die photo. Following that, they present a per-unit power estimation devised from performance counters. They train the submodel with a set of specialized micro-benchmarks to stress the correlated power units one by one. Many hardware events can reflect more power characteristics, making a contribution to the predicted accuracy. We treat things as a whole and combine different units on a higher level view to avoid a cumbersome per-unit model. In addition, our tool uses auto-calibration methodology that enables porting it to different processors without the need to recalibrate manually if, for example, the voltage–frequency curve is changed.

Shao and Brooks [30] propose to characterize workloads independent of the ISA. They profile memory, branch, and opcode behavior of an application using the intermediate representation of an application instead of the binary. Following up on this work, they propose using the ISA-independent profiles for speeding up the design of accelerators [31]. They achieve good speedups with minimal loss in accuracy both for performance and power estimations. In addition, Sam Van et al.

[32] demonstrated an analytical performance and power model, based on microarchitecture-independent application profiles to enable the evaluation of large design spaces using a single application-specific but architecture-independent profile. Performance and power are estimated with an average error of 13% and 7%, respectively.

Table 3.5 provides a high-level comparison of recent power estimation methods that are performance counter based and outlines the key differences among these methods when compared to our methods described in this work. The reference of each of the prior methods has the details.

3.7 Summary and Future Work

In this work, we proposed a new methodology that combines power, performance, and thermal measurements, together with selected performance counters to provide accurate low-level power estimations. The method was demonstrated on a modern out-of-order core including a breakdown to its key sub-domains: frontend, execution, and memory. A tool that implements the methodology is built and calibrated for the Skylake processor—newly released Intel. We demonstrated how the power estimation and per-domain breakdown reflect the expected behavior for a few handcrafted micro-benchmarks as well as when select software optimizations are used: vectorization, loop unrolling, and the new FMA instruction.

This work provides a detailed power and performance characterization breakdown for the SPEC CPU2006 benchmarks as measured on the Skylake system. While our detailed analysis substantiates the general trend that the core power correlates with performance (IPC), the details are important. For example, FLOP/cycle is a better FP efficiency metric than plain IPC—both metrics are included in the report. We have also found that the frontend consumes a sizable fraction of the core power even when it is not a performance limiter which hopefully suggests room for software and hardware optimizations.

For future work, we plan to extend the tool to model other processor cores which can help to evaluate and understand the efficiency of optimizations at a finer grain level. We also plan to enhance the tool to provide hints to software developers and guide the user to catch power bottlenecks in the programs similar to performance hints given by the Top-down Microarchitecture Analysis Method.

Appendix

The next table provides a power and performance characterization report for the SPEC CPU2006 benchmarks. The remaining power numbers are modeled by our tool. Note RAPL.PP0 accounts for off-core power as well (e.g., LLC), the RAPL.PPL0 will be close to the core power once the L2 cache miss is close to 0 and only one core is working. See Table 3.4 for system setup details. In addition, note that the sum core power breaks down to frontend, execution, memory, and global, the P_{Global} is constant which does not appear on the table, its value is about 0.6 W.

References

1. TOP 500 SUPERCOMPUTER SITES, http://www.top500.org/list/2013/06 (accessed December 12, 2013).
2. THE GREEN500 SITES, http://www.green500.org (accessed December 12, 2013).
3. Asanovic, K., Bodik, R., Demmel, J., Keaveny, T., Keutzer, K., Kubiatowicz, J., Morgan, N., Patterson, D., Sen, K., Wawrzynek, J. and Wessel, D., 2009. A view of the parallel computing landscape. Communications of the ACM, 52(10), pp. 56–67.
4. Haj-Yihia, Jawad, et al. "Compiler-directed power management for superscalars." ACM Transactions on Architecture and Code Optimization (TACO) 11.4 (2015): 48.
5. Merkel, A. and Bellosa, F., 2006, April. Balancing power consumption in multiprocessor systems. In ACM SIGOPS Operating Systems Review (Vol. 40, No. 4, pp. 403–414). ACM.
6. Bhattacharjee, A. and Martonosi, M., 2009, June. Thread criticality predictors for dynamic performance, power, and resource management in chip multiprocessors. In ACM SIGARCH Computer Architecture News (Vol. 37, No. 3, pp. 290–301). ACM.
7. Lee, K.J. and Skadron, K., 2005, April. Using performance counters for runtime temperature sensing in high-performance processors. In Parallel and Distributed Processing Symposium, 2005. Proceedings. 19th IEEE International (pp. 8-pp). IEEE.
8. Bircher, W.L. and John, L.K., 2007, April. Complete system power estimation: A trickle-down approach based on performance events. In Performance Analysis of Systems & Software, 2007. ISPASS 2007. IEEE International Symposium on (pp. 158–168). IEEE.
9. Intel Corporation, "Intel® 64 and IA-32 Architectures Optimization Reference Manual, Appendix B.1" Intel. (as of August 2014).
10. Isci, Canturk, et al. "An analysis of efficient multi-core global power management policies: Maximizing performance for a given power budget." Proceedings of the 39th annual IEEE/ACM international symposium on microarchitecture. IEEE Computer Society, 2006.
11. Rotem, Efraim, et al. "Power-management architecture of the intel microarchitecture code-named Sandy Bridge." IEEE Micro 32.2 (2012): 0020-27.
12. Rotem, Efraim, et al. "Energy Aware Race to Halt: A Down to EARtH Approach for Platform Energy Management." (2012): 1–1.
13. David, H., Gorbatov, E., Hanebutte, U.R., et al.: RAPL: memory power estimation and capping. In: 2010 ACM/IEEE International Symposium on Low-Power Electronics and Design (ISLPED), pp. 189–194. IEEE (2010).
14. Jawad Haj-Yihia, Ahmad Yasin, Yosi ben asher, Avi Mendelson, "Core Power breakdown tool", 2016.
15. Bhunia, S., Mukhopadhyay, S. (eds.): Low-Power Variation-Tolerant Design in Nanometer Silicon. Springer Verlag (2010).
16. Bellosa, F.: The benefits of event: driven energy accounting in power-sensitive systems. In: Proceedings of the 9th Workshop on ACM SIGOPS European Workshop: Beyond the PC: New Challenges for the Operating System, pp. 37–42. ACM (2000).
17. Powell, Michael D., et al. "CAMP: A technique to estimate per-structure power at run-time using a few simple parameters." High Performance Computer Architecture, 2009. HPCA 2009. IEEE 15th International Symposium on. IEEE, 2009.
18. Bertran, Ramon, et al. "Decomposable and responsive power models for multicore processors using performance counters." Proceedings of the 24th ACM International Conference on Supercomputing. ACM, 2010.

19. Bertran, Ramon, et al. "A systematic methodology to generate decomposable and responsive power models for CMPs." IEEE Transactions on Computers 62.7 (2013): 1289–1302.
20. A. Yasin, "A Top-Down Method for Performance Analysis and Counters Architecture," presented at the Performance Analysis of Systems and Software (ISPASS), 2014 IEEE International Symposi-um on, 2014.
21. Intel Corporation, "Intel open source", online: http://download.01.org/perfmon/ [accesses October 8, 2015].
22. Isci, C., Martonosi, M.: Runtime power monitoring in high-end processors: Methodology and empirical data. In: Proceedings of the 36th Annual IEEE/ACM International Symposium on Microarchitecture, p. 93. IEEE Computer Society (2003).
23. Singh, K., Bhadauria, M., McKee, S.A.: Real time power estimation and thread scheduling via performance counters. ACM SIGARCH Computer Architecture News 37(2), 46–55 (2009).
24. Intel® 64 and IA-32 Architectures Software Developer's Manual. Volume 3A: System Programming Guide, Part 1, [accesses January, 2016a].
25. Firasta, Nadeem, et al. "Intel avx: New frontiers in performance improvements and energy efficiency." Intel white paper (2008).
26. A. Carvalho, "The New Linux 'perf' tools," presented at the Linux Kongress, 2010.
27. A. Kleen, "toplev manual (pmu-tools)", online: https://github.com/andikleen/pmu-tools/wiki/toplev-manual [accesses October 8, 2015].
28. ThinkPad SMAPI kernel module version 0.40. http://tpctl.sourceforge.net/.
29. Bertran, R., Gonzàlez, M., Martorell, X., et al.: Counter-Based Power Modeling Methods: Top-Down vs. Bottom-Up. The Computer Journal 56(2), 198–213 (2013).
30. Y. S. Shao and D. Brooks, "ISA-independent workload characterization and its implications for specialized architectures," in Proceedings of the International Symposium on Performance Analysis of Systems and Software (ISPASS), 2013, pp. 245–255.
31. Y. S. Shao, B. Reagen, G.-Y. Wei, and D. Brooks, "Aladdin: A preRTL, power-performance accelerator simulator enabling large design space exploration of customized architectures," in Proceedings of the 41st Annual International Symposium on Computer Architecture (ISCA), 2014, pp. 97–108.
32. S. Van den Steen, S. De Pestel, M. Mechri, S. Eyerman, T. Carlson, L. Eeckhout, E. Hagersten, and D. Black-Schaffer. Micro-architecture independent analytical processor performance and power modeling. In Proceedings of the 2015 IEEE International Symposium on Performance Analysis of Systems and Software (ISPASS), Mar. 2015.
33. Intel Corporation, "6th Generation Intel® Processor Family – Specification update", online: http://www.intel.com/content/www/us/en/processors/core/desktop-6th-gen-core-family-spec-update.html [accesses August, 2016].
34. Spiliopoulos, Vasileios, Andreas Sembrant, and Stefanos Kaxiras. "Power-sleuth: A tool for investigating your program's power behavior." Modeling, Analysis & Simulation of Computer and Telecommunication Systems (MASCOTS), 2012 IEEE 20th International Symposium on. IEEE, 2012.
35. https://drive.google.com/open?id=0B3IgzCqRS5Q_ZGN0QVFqaWxxY28.

Chapter 4
Compiler-Directed Energy Efficiency

4.1 Introduction

The continuation of Moore's law allows the integration of increasing number of transistors onto a single die, and is expected to deliver higher transistor density for the foreseeable future. This increase in transistor count alongside the increase in processor frequency introduces demanding power delivery and energy challenges. Power delivery is becoming a first-order constraint for high-performance and energy-efficient systems [1].

Modern out-of-order processors contain complex structures in order to exploit Instruction Level Parallelism (ILP). Processors such as 2nd Generation Intel® Core™ [2] further add vector instructions that allow 256-bit wide data operations. These result in high-performance processors but introduce very high-power demands. The dynamic range of power, from the lowest activity levels of the processor, e.g., while waiting for data return from memory to the highest power required for simultaneous execution accessing all data ports with full-width data, can be very wide. This wide dynamic range is further extended by modern power management techniques such as Turbo [3]. Furthermore, these power transients can occur within a few core clock cycles, faster than the ability of existing control techniques to respond, which in turn cause instantaneous high-power excursions. Consequently, the power delivery network needs to be able to handle these power excursions by design.

Designing a system for power excursions at the worst-case workload and the highest possible frequency is impractical. It drives high system cost and is often infeasible. Such a design would require unacceptable performance compromises and would inflict power and performance penalties upon all workload periods that consume less than the worst-case power excursion.

In this study, we present a novel compiler-assisted power management method to overcome power excursions. We have modified the LLVM compiler [4] and have extended it with a power model to detect high-power code regions at compile time.

© Springer Nature Singapore Pte Ltd. 2018   107
J. Haj-Yahya et al., *Energy Efficient High Performance Processors*,
Computer Architecture and Design Methodologies,
https://doi.org/10.1007/978-981-10-8554-3_4

The compiler identifies high- and low-power phases in the source code, and encapsulates them with a short instrumentation code. This code emulates a new instruction, VEL (Voltage Emergency Level). This instruction should be interpreted as NOP on older processors. We have emulated the new instruction using a short sequence of instructions (five instructions and debug configuration) that trigger an internal power management event in the Intel® Core™ processor's Power Management Unit (PMU). This instrumentation code hints the hardware about potential high power. The hardware takes actions to protect against potential power excursions by either increasing the voltage guardband or lowering the frequency. The default state of the processor is high-power phase. Applications that have not been compiled with our compiler are still able to run at a higher power state without causing a malfunction. The compiled code unleashes the additional power head-room only to code regions that have been marked as low power. We evaluate the method on a high-end processor using the SPEC CPU2006 benchmark suite. We have used an offline simulator over trace data generated by the compiled benchmark runs on the target systems (both power-constrained and non-constrained systems). Using the simulator, we have measured up to 16% performance speedup on a power-constrained system and up to 11.4% energy savings on a non-constrained system. Compile time techniques have inherent limitations in predicting runtime behavior because the actual power consumption varies due to runtime dependencies such as input data, control flow, and micro-architectural profile. We have demonstrated on our system that these inherent limitations do not leave much unrealized gain. We have also validated the safety of the implementation and have identified no escapees that might compromise the execution.

This work makes the following contributions:

- We develop and implement a novel compiler-assisted hardware method to mitigate voltage emergencies.
- The proposed method requires minimal incremental changes, does not require widespread design methodologies or architecture changes, and is backward compatible.
- We validate the proposed method on the most recent Intel Core™ processor [5, 6] and measure promising performance speedup and energy savings using an offline simulator on the power trace data.
- We make the compiler power profiling tools available for the research community [7].

4.2 Power Delivery Constraints

High-performance processors may consume tens to hundreds of Amperes at sub-1V. This demand makes the Power Delivery Network (PDN) a highly constrained hardware resource both thermally and electrically.

4.2.1 Maximum Current Delivery

Voltage Regulators (VRs) suffer conversion losses primarily due to parasitic resistance on the power FET (Field-Effect Transistors) drivers and inductors, and due to gate capacitance of the FET switches. These losses translate into heat that might damage the VR components [1]. Heat develops relatively slow and allows control circuits to manage the power consumption [8–10] and is not the focus of this work. The maximum instantaneous current that can be delivered by a voltage regulator is also limited. The FET drivers may be damaged by high current and the inductors may reach magnetic saturation, causing the VR to malfunction. Over-current protection circuitry may turn off the VR when the maximum current is exceeded. These electrical limits occur much faster and are the focus of this study. The instantaneous high-power events can be handled by building the PDN for the worst case, even if it is rare [11]. If the VR cannot sustain the highest instantaneous power of the CPU ("Power delivery constrained system" in this work), the CPU needs to run at a lower voltage and frequency. In this work, we lower the frequency only for high-power intervals, hence gaining back this lost performance.

4.2.2 Voltage Droops

A simplified model of power delivery is described in Fig. 4.1. Power distribution systems are essentially resistive (R) and inductive (L) [12]. These parasitic components can cause AC and DC voltage droops that compromise processor's minimum or maximum supply voltage level [12, 13]. Voltage droops may be separated into static IR-drop (resistive noise) and dynamic $L \cdot \partial I / \partial t$-drop (inductive noise). The former is the static voltage drop due to the resistance of the power delivery network interconnects and is proportional to the DC impedance of the power delivery

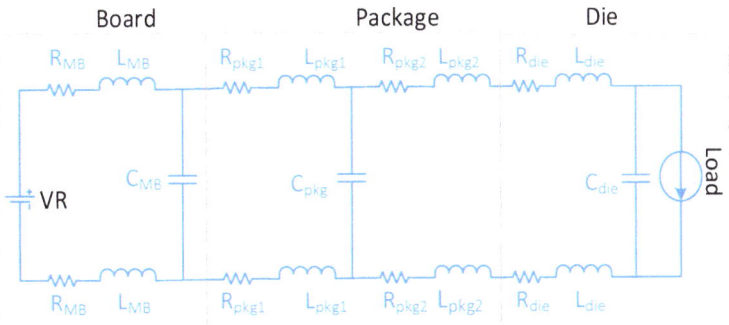

Fig. 4.1 Simplified RLC model for interconnections between the power supply and the load (processor)

network. The latter is caused by the inductance and the capacitance in the PDN and represents the transients of voltage noise when load current changes.

The power delivery system of a microprocessor ideally strives to maintain a low constant impedance across all frequencies. In practice, this necessitates several stages of decoupling to optimally flatten the supply impedance across a broad range of frequencies, as shown in the simplified circuit diagram in Fig. 4.1. Decoupling capacitors in each stage serve as local storage to supply charge to the next stage when needed quickly. For the core supply, it is generally impractical (in area and in cost) to place sufficient die capacitance to achieve near perfect filtering [1, 14]. A practical solution leads to several distinct resonances of the power supply impedance. When the processor transits from low-power state to high-power state in a few clock cycles, the increase in rate of current change ($\partial I/\partial t$) results in voltage droops due to resistive and inductive effects of the power distribution network. As shown in Fig. 4.4 [15], these voltage droops can be categorized into three distinct ones. These droops correspond to each stage of the decoupling capacitor present in the network. The first droop is influenced by the on-die capacitance and package inductance and typically occurs in a few nanoseconds time period. The second droop is influenced by the package capacitance and the socket inductance and usually occurs in a few tens of nanoseconds time period. The third droop typically occurs at hundreds of nanoseconds to a few microseconds time and is influenced by the motherboard capacitors, voltage regulator bandwidth, and the resistance of the PDN. The design goal is to minimize these voltage droops and to maintain low PDN impedance across wide frequency range to achieve maximum operating frequency.

The processor's manufacturer builds the package and die PDN and publishes specifications and design guidelines [16] for the external PDN to keep the impedance at target load line impedance (ZLL in Fig. 4.2). This study primarily addresses this external portion of the PDN, while assuming that the board has been designed according to manufacturer guidelines [16].

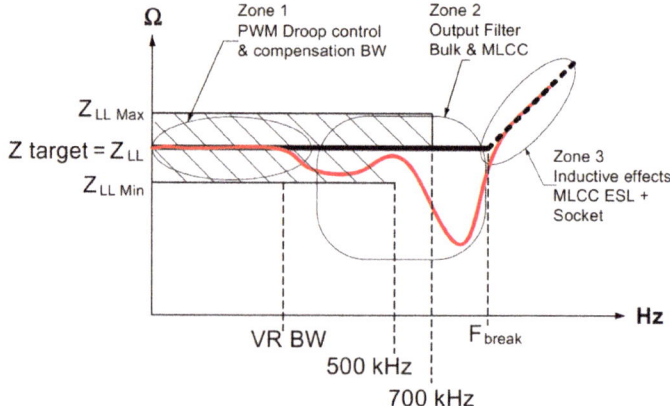

Fig. 4.2 Power distribution impedance versus frequency [16]

Short power (current) conjunctions are handled by the filter capacitor network on die and package. For a high-power (current) event to be observed by the board and VR, it needs to last hundreds of nanoseconds to a few microseconds (few hundreds to a few thousands of core clock cycles), depending on PDN design. With this observation, the VR and its connection to the processor are shown in the simplified model in Fig. 4.3. This model describes the load line or Adaptive Voltage Positioning (AVP) [16, 17] behavior as it appears to the VR and board. In this model, short current bursts (at the first and second droop frequencies) are filtered out by the decoupling capacitors while long current bursts (equal to or below the third droop frequency) are observed by the board and VR.

AVP keeps the load voltage close to V_{max} when the load current is low, while the load voltage will drop to close to V_{min} when the load current is at the maximum allowed level (I_{max}). In addition to cost reduction of the PDN [17], AVP allows

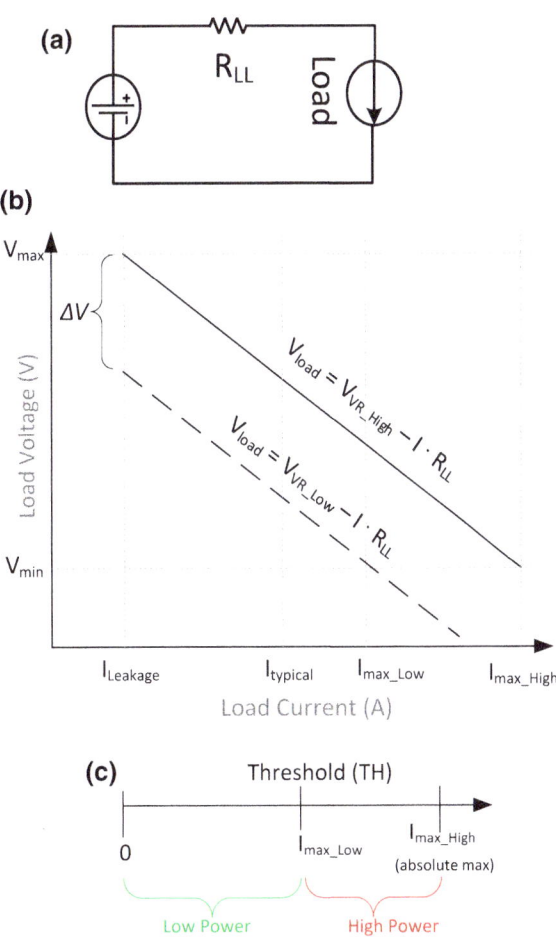

Fig. 4.3 a Simplified PDN model with load line, **b** load line with different maximum current levels, **c** low- and high-voltage guardbands based on threshold

reducing the power consumption at high loads by reducing the load voltage as shown in Fig. 4.3. The lowest allowable voltage V_{min} is determined by the maximum processor current (I_{max}) that can be drawn at a given frequency, as this I_{max} current determines the initial voltage guardband that compensates for voltage droop once this high current occurs. If we can limit or reduce I_{max}, then we will be able to reduce the voltage guardband to a lower voltage level for the same current. As shown in Fig. 4.3, the maximum current is I_{max_High}. If we can limit the maximum current to I_{max_Low}, then workloads with current between $I_{leakage}$ and I_{max_Low} can run with voltage lower by ΔV than the baseline voltage. This will save power consumption in proportion to the square of the load voltage, and in power-constrained modes we will be able to use this "freed" power budget to raise processor frequency and to gain higher performance relative to the baseline. In this study, we characterize program code regions based on the maximum current that can be drawn. This is done using compiler and power model as shown in Sect. 4.4.

We focus on the third voltage droop while assuming that the first and second droops are handled by the on-die and package decoupling capacitors, load line-based voltage optimizations are done by the processor, in addition to adding voltage guardband at manufacturing time. Some previous studies have also addressed these effects [18–23].

A voltage regulator that can functionally support instantaneous high current (referred to as "unconstrained system" in this study) still needs to drive a higher steady state voltage, which causes square cost in energy. In the unconstrained system scenario of this study, the processor runs at the highest frequency. While during high-power phases, when current excursions might cause a voltage droop, the voltage needs to be increased, and at low-power phases a lower voltage can be maintained. The increased energy is consumed only in the high-power phases, resulting in energy savings compared to a non-protected system that consumes increased energy for the entire runtime.

4.2.3   Voltage Emergencies Prediction

Several previous studies address voltage emergencies prediction [24–26] for different types of voltage droops. In the following, we explain our method of detecting voltage emergencies using compiler and power model (Fig. 4.4).

As explained above, this study focuses on the third droops and VR maximum current violation. To observe third droops, high-power burst over a relatively long execution window should be generated. This burst discharges the decoupling capacitors network on die and on package, and the charge stored on board capacitors starts to be used (the VR is not responding at this stage as the burst is faster than its bandwidth). Consequently, we observe a voltage droop at the load voltage, as shown in Fig. 4.5. This droop is affected mainly by PDN resistance, as high current flows into the processor load line (from board capacitors to processor), causing high-voltage droop (IR-drop). During system design, additional voltage

Fig. 4.4 First, second, and third droops in the time domain [42]

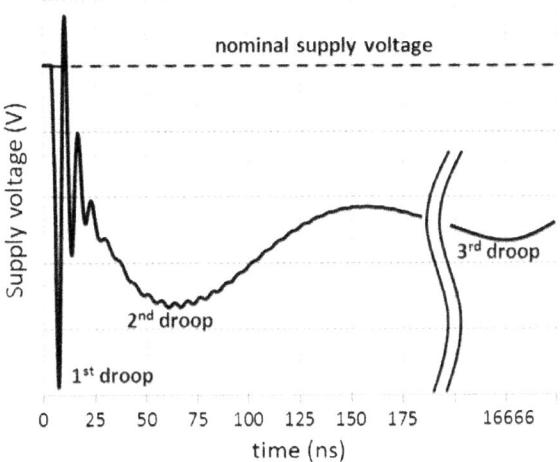

guardband is added to nominal voltage to prevent dropping below minimum operation voltage when such a burst arrives. The guardband width is relative to the maximum current that can be drawn by the processor.

Figure 4.5 provides intuition into the behavior of voltage as seen by the board and VR while executing high-power instruction over short- and long-time intervals. We can see that the short burst of instruction execution causes the voltage to drop slightly. This burst is sufficiently short so that the network begins to recover before the minimum operation voltage limit is crossed, due to relatively low current consumption from the board capacitors. The package capacitor stores sufficient charge to satisfy this burst, and the low current from the board capacitors is used to recharge the package capacitors. In the case of a longer burst, voltage drops below the minimum operation voltage limit; in this case, a higher voltage guardband is needed.

To predict a third droop voltage emergency, we predict the maximum current that can be drawn over a given instruction window. For code regions that consume high power (current), our framework indicates a higher voltage guardband, while

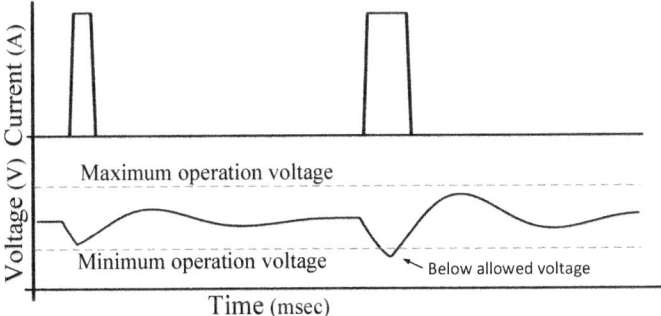

Fig. 4.5 Voltage droops relative to load

for relatively low-power (current) code regions, we reduce the voltage guardband, as shown in Fig. 4.3b. To determine the high-power code regions, we use a power model (discussed in Sect. 4.4.2). With this power model, we estimate the overall energy consumed by a fixed length window of instructions and classify code regions power/current levels by comparing this energy to an energy threshold.

The energy consumed by a fixed window is correlated to current as follows. Energy is $E = P \cdot T$. Time T is assumed (the length of the instruction window) and power is $P = V \cdot I$. Voltage V is also assumed constant, set by the processor PMU for the entire instruction window. Thus, the total energy E consumed by the fixed instruction window is correlated to the current I. The length of the instruction window is chosen to be close to the inverse of the resonant frequency of the third droop of the processor (hundreds of nanosecond to few microsecond). For our system, a window of 500 instruction has been used.

Based on this observation, voltage emergency can potentially happen if the total energy consumed by an instruction window exceeds an energy threshold TH.

4.3 The Algorithmic Problem

Following the observation in Sect. 4.2.3, the solution for the problem of voltage emergencies can be mapped to solving an algorithmic problem on the Control Flow Graph (CFG) of the source code. The algorithm objective is to mark safe and unsafe code regions on the CFG. Safe code region is code that does not cause voltage emergencies or maximum current violations when executed, while unsafe code region is code that might cause voltage emergencies or maximum current violations. Unsafe code must run at higher voltage or lower frequency to preserve processor execution correctness (as discussed in Sects. 4.2.1 and 4.2.2).

To predict safe code regions, the algorithm assures that a given instruction window of K instructions does not consume total energy that exceeds an energy threshold TH. If that threshold is exceeded by some code region, then that code region is marked with "+" (must run at higher voltage or reduced frequency). Otherwise, the code is marked with "−" (can run with nominal voltage and nominal frequency). A CFG with unsafe code regions marked with "+" and safe code regions marked with "−" is defined as *K-TH legal*.

4.3.1 Problem Formal Definition

Given a directed graph G with cycles (the CFG) such that

- G has a start node s with a path to every other node v,
- All nodes have weights (Energy Per Instruction), then a *power assignment* to G is a labeling of some nodes by "+" (start of high-power phase) and some nodes by "−" (start of low-power phase).

We define the following:

- Let $P_k = v_1 \rightarrow v_2 \rightarrow \dots \rightarrow v_k$ be a path of length k, possibly with cycles.
- A node $v \in G$ is under the influence of "+" if all paths from s to v contain a node marked with "+" that is not overridden by a "−" node.
- A node $v \in G$ is under the influence of "−" if there is some path from s to v that contains a node marked with "−" that is not overridden by a "+" node.
- A power assignment to G is **K-TH legal** if all paths $P_k = v_1 \rightarrow \dots \rightarrow v_k$ of length $k \leq K$ with total weights $\geq$ TH have their first node v_1 under the influence of "+" and the rest of P_k nodes $v_2,..,v_k$ are not labeled by "−".
- The *profit* of a K-TH legal power assignment is the total length of paths with length $k >$ K and total weights $<$ TH that are under the influence of "−".

Given G as above, we seek to find K-TH power assignment with maximal profit (i.e., maximize the number of instructions that are labeled as low power, and hence can be executed with low-voltage or non-reduced frequency).

4.3.2 K-TH Legal Graph Example

Consider the graphs in Fig. 4.6 that represents subgraph of a CFG of some program. The nodes represent instructions and the number near a node represents the weight of the instruction. For K = 3 and TH = 4, these graphs have an optimal assignment with the labeling ("+" and "−") shown.

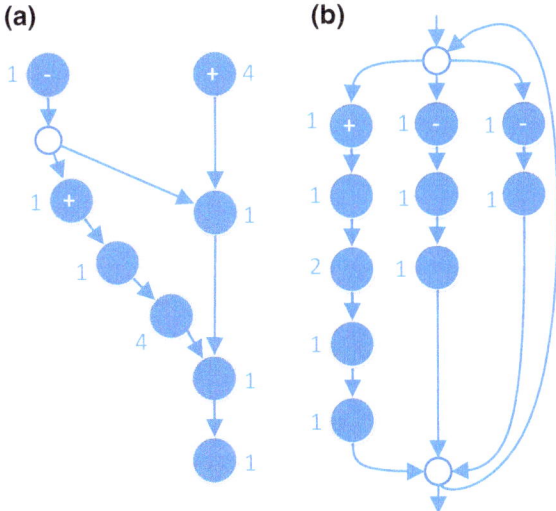

Fig. 4.6 Examples of optimal power assignments when K = 3 TH = 4 for **a** three paths graph, **b** loop with three paths

4.3.3   The Algorithm

We first define the linear solution for the special case that G is a path L of size
n > K.

1. Let sum_k (v) be the total sum of the weights of v and the next K-1 nodes
   following v.
2. Scan path L in topological order. For each v along the scan:
3. if sum_k (v) $\geq$ T then
4. If v is not labeled with Red then label v by a '+'
5. Label K-1 successors of v with Red, and remove any '−'
6. Label the Kth successor of v with '−'

   The proposed (nonoptimized) algorithm works as follows:

1. Start with CFG of a function.
2. Label all nodes with Blue.
3. Unroll each loop enough many times until all possible paths inside the loop
   body are exposed and the shortest path is of length 2·K. Let G be the outcome of
   this unrolling with a unique start node s and an end node t.
4. Let *cover*(G) be the set of all paths from s to t that do not pass through the same
   edge more than once.
5. For each path R $\in$ *cover*(G), we apply the linear solution labeling some of G
   nodes with "+" or "−".
6. Replace CFG with the labeled graph G.
7. Before an instruction labeled with "+", insert an instruction that hints to the
   hardware of an entry to high-power code region (see Sect. 4.4.1 for a description
   of the VEL instruction).
8. Before an instruction labeled with "−", insert an instruction that hints to the
   hardware of an entry to low-power code region.

4.3.4   Algorithm Description and Example

The algorithm objective is to classify code regions into two group, high-power
(current) and low-power (current) regions based on a threshold. For a high-power
(current) burst to be observed by the board or VR, it needs to last a few hundreds of
nanoseconds to a few microseconds at least; a short burst is handled by the die and
package decoupling capacitors (as describes in Sect. 4.2.2).

Consider a sequence of K instructions, where K is chosen as the number of
cycles needed for a high current burst to draw a third droop. We calculate the
energy consumption of each instruction (see Sect. 4.4.2). For example, a scalar
move (mov) instruction consumes less energy than a vector move (vmovups)
instruction. We then estimate the total energy consumed by the instruction
sequence. If the total energy exceeds a threshold TH, then we mark the sequence as

high power. This is achieved by inserting a VEL 1 instruction (described at Sect. 4.4.1) in the beginning of the sequence and a VEL 0 at the end. In case of an instruction path longer than K, this process is applied to each subsequence of length K of the path (this is defined as "linear solution" in the algorithm of Sect. 4.3.3). VEL is a per-thread indication that indicates the voltage emergency level of the subsequent code arriving at the processor's PMU.

One of the algorithm's challenges is to figure out all high-power code sequences (code sequences of length K whose total energy exceeds the threshold TH). This can be done by traversing the code CFG and searching for high-power paths of length K. We also need to consider paths that iterate over loop body (assuming loop body < K); to expose such paths, we use a nonoptimal solution, by unrolling loops enough many times to discover all possible paths of length K that can start at any point in the loop.

Once loop unrolling is done, the algorithm traverses all paths of each function, starting from the entry basic block and proceeding until the exit basic block. The linear solution is applied to each such unique path.

The algorithm is exemplified on a code snippet taken from the 433.milc benchmark of the SPEC CPU2006 benchmark suite [27]. The code snippet is shown in Fig. 4.7. The benchmark has been compiled with the LLVM compiler using –O3 flag (Auto-vectorization enabled by default), tuned for "corei7-avx" (for AVX2 instruction set, [28]).

;;Code snippet from 433.milc of SPEC2006 benchmarks			Normalized MEPI
	...		
.LBB44_66:	movq	$-576, %rdi	5.2
	.align	16, 0x90	
.LBB44_67:			
	movq	%rsi, 688(%rdx,%rdi)	5.2
	vmovups	%xmm0, 736(%rdx,%rdi)	28.6
	vmovups	%xmm0, 720(%rdx,%rdi)	28.6
	vmovups	%xmm0, 712(%rdx,%rdi)	28.6
	vmovups	%xmm0, 696(%rdx,%rdi)	28.6
	movq	%rsi, 752(%rdx,%rdi)	5.2
	vmovups	%xmm0, 800(%rdx,%rdi)	28.6
	vmovups	%xmm0, 784(%rdx,%rdi)	28.6
	vmovups	%xmm0, 776(%rdx,%rdi)	28.6
	vmovups	%xmm0, 760(%rdx,%rdi)	28.6
	movq	%rsi, 816(%rdx,%rdi)	5.2
	movq	$0, 824(%rdx,%rdi)	5.2
	addq	$144, %rdi	3.8
	jne	.LBB44_67	1
.LBB44_68:			
	addq	$2048, %rdx	3.8
	incl	%ecx	2.1
	cmpl	%eax, %ecx	2.2
	...		

Fig. 4.7 Code snippet from 433.milc benchmark of SPEC06

Fig. 4.8 **a** CFG of the code snippet, **b** CFG with loop unrolled

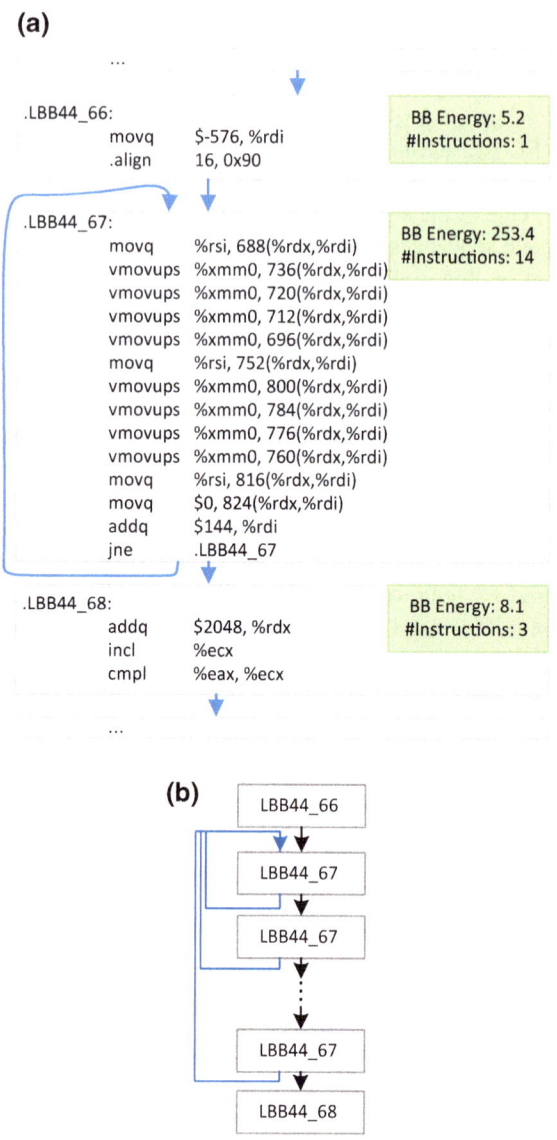

For every instruction, Fig. 4.7 shows the Normalized Maximum Energy Per Instruction (Normalized MEPI). It represents the weight of the instruction and estimates the maximum energy that can be consumed by executing the instruction. Calculating Normalized MEPI is described in Sect. 4.4.2.

Figure 4.8 shows the CFG of the code snippet before and after loop unrolling. The upper right-hand side of each basic block indicates the total energy (BB Energy) of the basic block and the number of instructions at the basic block. We can see that basic block "LBB44_67" (the loop body) consumes much higher energy relative to the other two basic blocks.

For K = 500 (instructions window) and TH = 9000 (Energy threshold), after unrolling the loop body (LBB44_67) 36 times, we observe that the unrolled loop body has 14 × 36 = 504 instruction and its energy is 253.4 × 36 = 9122, higher than the threshold TH. Consequently, VEL 1 is inserted at the loop entry to indicate a high-power (current) loop, and VEL 0 is inserted at the end (beginning of LBB44_68).

From this example, we observe that the high-power event within the window of 500 instructions is caused mainly by the 128-bit vector instructions (e.g., vmovups). If we replace each such instruction with a 64-bit instruction (e.g., replacing 128-bit "mov" by two 64-bit "mov"), we will at least double the number of instructions at the loop body while each instruction consumes approximately half the power; this replacement eliminates the high-power event, but performance is reduced (taking more cycles to perform the same task).

4.4   Framework

In order to mitigate voltage emergencies and maximum current violation problem in our processor, we have created a framework comprising the following parts:

- VEL instruction emulation,
- Power model,
- LLVM compiler, and
- Voltage emergencies detection algorithm.

The high-level flow of the framework is shown in Fig. 4.9. The program is compiled with our modified compiler, using a Power Model to calculate the regions in the generated code that should be protected against voltage emergencies. The compiler inserts ("instruments") the new Voltage Emergency Level (VEL) instruction at the beginning and the end of the region with appropriate parameters.

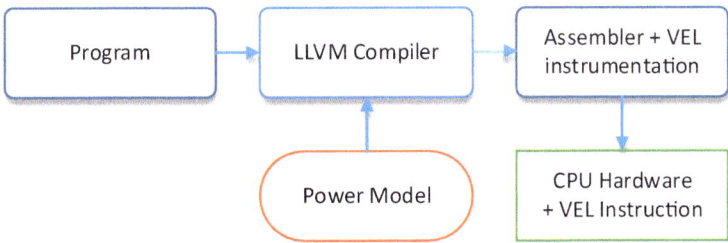

Fig. 4.9 Framework: compiler, Power Model, and VEL

4.4.1   VEL Instruction

The VEL instruction is designed to generate a hint from the software to the hardware. The instruction takes a floating-point operand which hints the level of voltage emergency that might be drawn by subsequent code. We define the voltage emergency level parameter as a fraction; 0 means no voltage emergencies are expected (low-power code), while 1 means that a voltage emergency is expected to happen after executing the code following the VEL instruction (high-power code). A value between 0 and 1 determines the code power level relative to high-power code that causes voltage emergency. In this study, we only use the values 0 and 1.

The hardware checks if the emergency level reaches 1. When this level is detected, the hardware can trigger the following actions to prevent voltage emergency:

1. If possible, raise voltage to a safe level corresponding to the voltage emergency level.
2. If the voltage cannot be raised, e.g., due to exceeding maximum operation voltage, lower the CPU frequency to a safe level.
3. Throttle the CPU frontend until the voltage or the frequency reach the safe level.

If the hint is 0, then the hardware can reduce voltage and increase frequency back to nominal levels.

The VEL instruction is stored per thread, allowing the hardware to predict voltage emergencies across a multithreaded system. With simultaneous multi-threading (SMT) or multi-core, each software thread sets its own VEL values. The hardware sums VEL values of all running threads and determines if a voltage emergency is expected. Although the proposed method takes multithreading into account, we focus on single-thread workloads in this study and leave multithreading for future work. Multi-core is discussed further in Sect. 4.6.

Implementing VEL as processor hardware is infeasible in this study. Instead, we emulate the VEL instruction, by employing instrumentation code and debug knobs of the processor. Once the instrumentation code is executed under debug configuration, the CPU core sends a special internal event to the PMU and reports this event at the Trace-Port (debug port) as shown in Fig. 4.10. The PMU raises the voltage if VEL code is 1 and reduces voltage back to nominal level when VEL code is 0. The trace data is used later by the simulator that reports power and performance gain based on VEL indications to the PMU.

4.4.2   Power Model

To determine if a given code segment can produce voltage emergency, we should be able to estimate the maximum power of this code. For this purpose, our model indicates Maximum Energy Per Instruction (MEPI). The energy absolute values

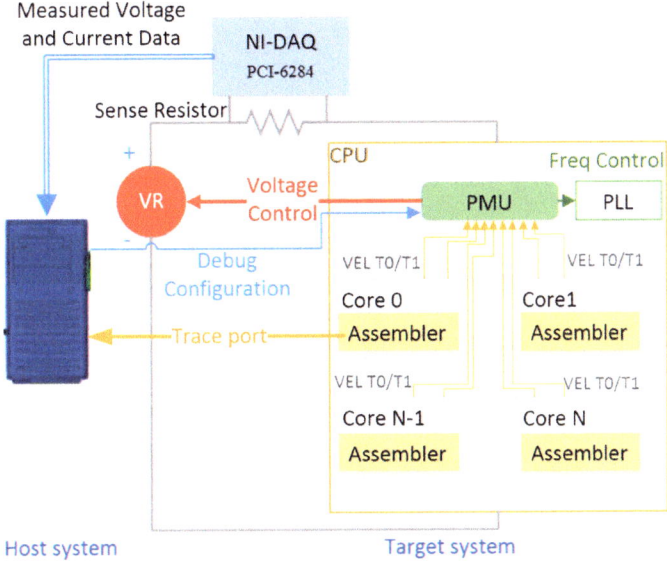

Fig. 4.10 VEL emulation flow description

depend on frequency, voltage level, temperature and fabrication process. For our purpose, we maintain Normalized MEPI, such that the instruction with minimal MEPI takes value of 1 and all other instructions are ranked relative to it. To measure MEPI, we have used technique similar to Shao et al. [29]. The idea is to develop a micro-benchmark that consists of a loop that iterates the same instruction numerous times. For power measurement, we have used a CPU energy counter [30]. This measurement is repeated many times while randomizing the instruction's address and data operands. A pseudocode for measuring MEPI is shown in Fig. 4.11.

```
set MEPI = 0
                        for 1 to numberOfTimeToRandomize
    Randomize instruction parameters (operands and data)
    energyBefore = ReadEnergyCounter()
    for 1 to Repeat
        execute: Instruction Op1, Op2, Op3
    end
    energyAfrer = ReadEnergyCounter()
        CurrentMEPI = (energyAfrer ó energyBefore)/ Repeat
    MEPI = Max (MEPI, CurrentMEPI)
end
```

Fig. 4.11 Pseudocode for measuring MEPI

Table 4.1 Part of Haswell CPU instructions Normalized MEPI

Instruction type	Description	Normalized MEPI
FMA256	Fused multiply–add 256bit	98.2
Store256	Vector store of 256bit	87.8
Load256	Vector load of 256bit	70.8
Store128	Vector store of 128bit	59.1
Load128	Vector load of 128bit	50.8
FMA128	Fused multiply–add 128bit	48.8
FMUL128	Floating-point multiply 128bit	38.0
FADD128	Floating-point add of 128bit	33.9
IMUL64	Integer multiply of 64bit	10.8
IMUL32	Integer multiply of 32bit	5.7
IADD32	Integer add of 32bit	2.1
MOV32	Registers move of 32bit	1

We have applied this method to our target processor and have measured MEPI for each instruction. We then normalized the MEPI values relative to the instruction with the minimal MEPI as shown in Table 4.1. In our target processor, the memory subsystem and caches are not sharing the same power supply with the cores; thus, MEPI values represent only the energy consumed from the core power supply.

4.4.3   LLVM Compiler

We used the open-source LLVM compiler [4] version 3.4. Figure 4.12 shows the LLVM block diagram. Compiler changes were made to the backend.

For our study, two main changes were made to the compiler:

4.4.3.1   Power Model Insertion to LLVM

The LLVM code generator uses the Target Description files (.td files) that contain detailed description of the target architecture. We added a new field for MEPI. Each type of instruction was mapped to its relevant MEPI. We have inserted the Normalized MEPI values for the X86 target as measured in Sect. 4.4.2.

4.4.3.2   Code Generator Pass

We have implemented a new machine function LLVM pass. The pass was inserted into the "Late Machine Code Opts" stage as shown in Fig. 4.12. The pass

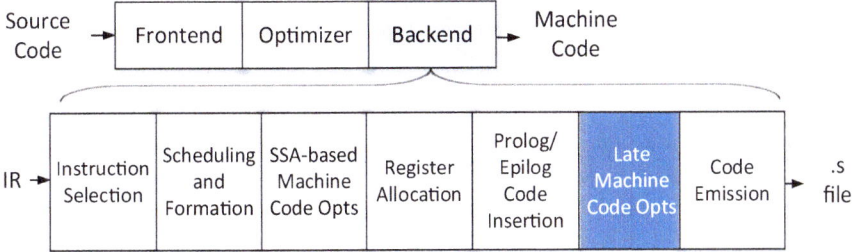

Fig. 4.12 LLVM block diagram

implements an algorithm for detecting code regions with potential voltage emergencies. The pass works on the machine-code CFG and uses the power model. The algorithm is described in Sect. 4.4.4.

4.4.4 Detection Algorithm

We apply a simplified variant of the algorithm described in Sect. 4.3. The simplified algorithm does not find the optimal profit, but it keeps code size similar to original code. The simplified algorithm works as follows:

1. Start with CFG of a function.
2. Duplicate CFG into G. Unroll each loop several times until all possible paths inside the loop body are exposed and the shortest path is of length 2·K.
3. Let cover(G) be the set of all paths from s to t that do not pass through the same edge more than once.
4. For each path R ∈ cover(G) apply the linear solution, labeling some of G nodes with "+" or "−".
5. For each loop LP in G, if LP contains a node marked with "+" then go to the original graph CFG and mark the pre-header of LP with "+" and the exit nodes with "−".
6. For all paths outside loops, apply the linear solution.

The algorithm outputs all instructions that were labeled by "+" or "−". Apply the following to labeled instructions:

- Before an instruction labeled with "+", insert VEL 1 instruction.
- Before an instruction labeled with "−", insert VEL 0 instruction.

4.5 Results

4.5.1 System Under Evaluation

The method is experimented on a platform that contains two systems, target system and host system, as shown in Fig. 4.10. The target system is the computer that runs the benchmark, containing 4th Generation Intel® Core™ processor i7 code name Haswell 4900MQ. The host system is a computer used to collect measurement data. The target system has been equipped with a National Instruments data acquisition (PCI-6284) connected to the host system for data collection. A debug port (Trace-Port) is connected from target to host. Through this port, the host collects the VEL instruction events, system-on-chip components power, and workload performance scalability with frequency (value between 0 and 1 which is defined as the percentage of performance increase over the percentage of frequency increase). Sampling of voltage, current, and Trace-Port data is carried out at a rate of once per one millisecond. A subset of SPEC CPU2006 benchmarks [27] has been used for power and performance measurements. Benchmark scores are the metric of performance.

The SPEC benchmarks have been compiled with the modified LLVM compiler with –O3 flag (Auto-vectorization enabled by default), tuned for "corei7-avx" (for AVX2 instruction set [28]).

The parameters for the detection algorithm, K and TH, have been determined using a search method. We have divided the instructions into two groups based on their MEPI. We search for the voltage level that allows 70% lower power instructions to pass without voltage emergencies assuming executing each instruction in an infinite sequence. Once this voltage level is determined, we check the upper 30% group of the instructions. We run the instruction with the lowest MEPI (that causes voltage emergency) in a sequence. The length of shortest sequence that still causes voltage emergency is K, and TH is the energy consumed by that sequence.

The modified LLVM compiler generates the code including instrumented code for VEL instruction emulation. Compilation time is increased by 8% on average relative to baseline due to long time for the detection algorithm. The instrumentation code is five instructions long and has no impact on actual benchmark performance. We have run all benchmarks with core frequency of 2500 MHz. A plot of the maximum power of each phase together with the VEL marker state (Fig. 4.13, where the smaller graph is a zoom-in) demonstrates how high-power phases are marked by our compiler.

We have created an offline simulator that scans through the captured traces and applies power management policy (i.e., frequency and voltage change) to each phase. Increased voltage and frequency result in increased power and shorter runtime of the interval. We have used Haswell power performance characteristics for power calculations, frequency transition cost, and the actual benchmark performance scalability with frequency.

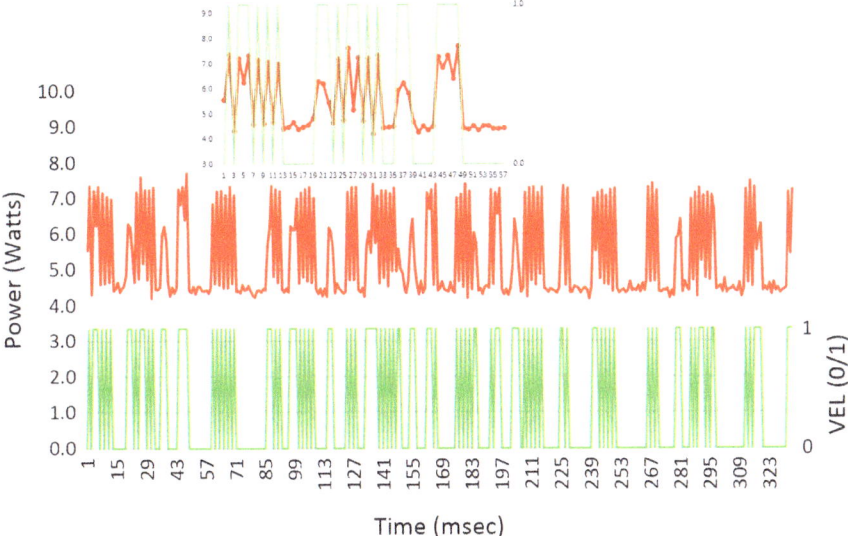

Fig. 4.13 Power trace and VEL marker for 464.h264ref run

4.5.2 Scenarios Evaluation

Two scenarios have been evaluated:

4.5.2.1 Power Delivery Constrained System

The workload is limited by instantaneous current. As a result, it needs to run at a lower frequency that guarantees safe operation. The compiler marks safe intervals where the processor can run at higher frequency and performance ("Performance gain" column in Table 4.2).

We observe that 75% of the benchmarks do not experience high-power excursion risk and can run at a higher frequency for the entire runtime. The most gaining benchmarks have frequency-sensitive bottlenecks as classified by top-down analysis [31]. For instance, 456.hmmer and 462.libquantum are "core bound", meaning that they are limited by the throughput of the core execution units. 445.gobmk, 458. sjeng, and 473.astar suffer much due to recovery from miss-predicted branches (how fast the frontend can fetch a corrected path is frequency sensitive when the instruction set is cache resident). The rest of the workloads gain performance only during safe intervals. The weighted average performance gain is 12.5%.

4.5.2.2 Non-constrained System

The power delivery network can supply high current excursions but the voltage has to be increased in order to compensate for voltage droops over the serial resistance. This contributes to increased energy consumption ("Energy savings" column in Table 4.2). Weighted average of 9.7% with up to 11.4% energy saving is achieved by lowering voltage during the safe intervals.

4.5.3 Technique Accuracy

Our method identifies potential power excursions at compile time. The actual power consumption is a function of runtime behavior, in particular data dependencies, control flow, and stalls due to memory access patterns. This means that code region marked by the compiler as high power may not draw high power due to actual parameters at runtime. For example, when one of the arguments of the multiply instruction (mul) is zero at runtime, it consumes much less power than expected by the compiler. The compiler uses worst-case power model on instructions (MEPI).

Two types of incorrect predictions can occur: False positive happens when we mark high-power phase while the actual runtime power is low. False negative happens when a high-power event is missed. False negative is critical because it can allow power excursions while the voltage is not configured for high power, possibly leading to runtime errors. We have scanned the power traces and did not identify

Table 4.2 Benchmark run results

Name	Time protected %	Performance gain %	Energy savings %
464.h264ref	99.6	0.8	0.3
403.gcc	39.0	9.1	6.7
447.dealII	24.8	10.7	8.3
470.lbm	0.0	12.0	11.4
433.milc	0.0	13.7	11.1
429.mcf	0.0	14.0	11.0
444.namd	0.0	14.0	10.9
483.xalancbmk	8.3	14.4	10.3
471.omnetpp	0.0	14.7	11.0
450.soplex	0.0	15.1	11.3
458.sjeng	0.0	15.2	11.0
462.libquantum	0.0	15.6	11.4
445.gobmk	0.0	15.8	10.9
473.astar	0.0	16.0	11.0
456.hmmer	0.0	16.0	11.1
Total	18.0	12.5	9.7

any such error in our test suite. It seems that false negative accuracy of our technique is 100%.

False positive is a noncritical event and translates into less than perfect gain. Scanning through the power traces, we have verified that all phases with high-power marking contain at least one high-power sample. Within these marked high-power phases, we identified 5.9% samples (1.1% of the total runtime) that consume low power. Hence, the accuracy of our technique is 94.1%.

4.6 Related Work

Hardware techniques: Researchers have focused on hardware mechanisms to characterize, detect, and eliminate voltage droops [11, 32, 33]. While these solutions have been effective at reducing $\partial I/\partial t$ [32] to the operating range of the processor, the executing program incurs performance penalties as a result. The hardware solutions are based on voltage control mechanisms that detect soft threshold violation by the processor and trigger fast throttling mechanism for the processor to reduce the $\partial I/\partial t$ effect. The hardware mechanism makes sure that voltage will not reach hard emergency voltage violation, and hence there will be cases of false alarms of the hardware mechanism. Other architectural techniques utilize some type of detection and recovery mechanism to deal with errors [22, 23, 34] and use redundant structures or replay mechanisms to detect and correct errors. All these techniques incur additional complexity or hardware overhead.

Some researchers explored detecting and mitigating errors via circuit techniques [35, 36]. The research using Razor systems assumes that errors will occur and insert redundancy within latches. Although effective, Razor requires significant new hardware and a completely different design methodology that fundamentally changes the way processors are designed.

Our work uses a relatively simple hardware mechanism, and the tuning process is relatively shorter than other methods discussed above. In addition, for detecting the third droops, the compiler approach provides a much more visible window relative to hardware mechanisms for detecting potential voltage droops.

Software and Compiler: A software approach to mitigating voltage emergencies was proposed by Gupta et al. [37]. They observe that a few loops in SPEC benchmarks are responsible for the majority of emergencies in superscalar processors. Their solution involves a set of compiler-based optimizations that reduce or eliminate architectural events likely to lead to emergencies such as cache or TLB misses and other long-latency stalls. Reddi et al. [38] proposed a dynamic scheduling workflow based on a checkpoint-and-recovery mechanism to suppress voltage emergencies. Once a code part causes a voltage margin violation, it is registered as a hotspot, and NOP injection and/or code rescheduling is conducted by the dynamic compiler. This flow is independent of architecture or workload. However, users should choose initial voltage margin properly in order to limit the rate of voltage emergencies. Reddi et al. [39] evaluate voltage droops in an existing dual-core CPU. They propose designing voltage margins for typical instead of worst-case behavior, relying on resilience mechanisms to recover from occasional errors. They also propose co-scheduling threads with complementary noise behavior, to reduce voltage droops.

Some researches discussed the impact of compiler optimization on voltage variations. Kanev et al. [19] showed that compiler-optimized code experienced a greater number of voltage droops, and, in certain cases, the magnitude of the droops was also noticeably larger. In a resilient processor design, this can eventually lead to performance loss for the more aggressively optimized case. In that work, the authors used a 45 nm chip that contained only 3% of the original package decoupling capacitor to imitate voltage droops at modern 22 nm processors. That work focused on first and second droops, while our work, although we also address the compiler, does not optimize the code but rather adds hinting instructions and focuses on the third droop.

Toburen [26] presented compilation techniques to mitigate the voltage fluctuations on VLIW architecture. They proposed a complier scheduling algorithm to eliminate the current spikes resulting from parallel execution of instruction on high-energy function units during program execution by limiting the amount of energy which can be dissipated in the processor during any one core cycle. That method targeted the high- and mid-frequency voltage droops, while our work targets the third droop. Further, their method is suitable for VLIW architecture, while for superscalar out-of-order architecture the scheduling at compile level affects the execution order at the processor to a lesser degree.

Multi-core: As most of today's systems have multi-core processors, and in most of these processors the cores share the same power delivery network, increasingly, one core can either constructively or destructively interfere with activity of the other cores [18]. Constructive interference is bad because it amplifies voltage variation, while destructive interference is good because it dampens voltage variation. Reddi et al. [40] measured and analyzed droops on a two-core Intel system and discussed constructive and destructive interference between processors and the difference in droops between average and worst-case scenarios. This information was used to design a noise-aware thread scheduler to mitigate some of the $\partial I/\partial t$ stresses in the system.

Miller et al. [18] showed that multithreaded programs such as those in the PARSEC suite have synchronization points that could align the threads and produce opportunities for high $\partial I/\partial t$ stress. They used fluctuations in average power estimated (Intel RAPL interface [41]) at intervals of 1 ms on hardware as a proxy for expected $\partial I/\partial t$ variations. This may have captured third droop excitations. They also observed that barriers could cause destructive core-to-core interference during the execution of multithreaded applications. Their work eliminated voltage emergencies by staggering threads into a barrier and sequentially stepping over it. Our work predicts the voltage variation based on average energy of assembly instruction over a known interval. We rely on the PMU to handle the alignment cases by setting the appropriate voltage level based on the number of cores having high-power event.

Kim et al. [15] measured and analyzed $\partial I/\partial t$ issues on multi-core system. They built a tool to develop and automate $\partial I/\partial t$ stress-mark generation framework. They consider first and second droops that can occur in a multi-core, and showed that alignment occurred relatively often when threads consisted of short execution loops. Our work focuses on the third droop and maximum current violation.

More recently, Lefurgy et al. [21] addressed active monitoring and managing of the voltage guardband based on the use of a Critical Path Monitor (CPM). The CPM monitors the critical pathways in the processor and increases the voltage guardband if the CPM detects potential emergencies. Although CPM is a very effective mechanism, it requires additional hardware, monitoring mechanisms, and tuning of the CPM to detect and correct possible errors. In addition, that technique involves many false alarms as it looks at a narrow window of execution cycles to predict the third droop, while third-level droops develop at hundreds to thousands of cycles. Our method, on the other hand, considers a wider window of instructions as it is done at the software level of the compiler.

Voltage emergency prediction: For voltage emergency prediction, Reddi et al. [24] proposed a solution for eliminating emergencies in single-core CPUs. They employed heuristics and a learning mechanism to predict voltage emergencies from architectural events. Based on the signature of these events, they predicted potential voltage emergencies and showed that with signature size of 64 entries they were able to reach 99% accuracy. When an emergency was predicted, execution rate was throttled, reducing the slope of current changes. That method is good for predicting first and second droops as it looks at a short window of execution cycles (a few nanoseconds to a few tens of nanoseconds), while our approach predicts third voltage droops. As we work at the compiler level, we are able to look forward at hundreds of cycles ahead. This yields higher accuracy for predicting third droop relative to hardware solutions with a narrower window that look at the beginning of a sequence of instructions that might cause a droop.

Joseph et al. [25] proposed a control technique to eliminate voltage emergencies. The technique is based on a sensing mechanism at the circuit level that feeds control actuator. The actuator temporarily suspends the processor's normal operation and performs some set of tasks to quickly raise or lower the voltage back to a safe level. This work is using a circuit mechanism to detect voltage emergencies. It may be accurate for first and second droops but for third voltage droop it is not accurate, because third droop frequency is slow (hundreds of nanoseconds to few microseconds).

4.7 Multi-core and Multithreads Handling

Our work predicts the voltage variation based on average energy of assembly instruction over fixed interval. This method estimates the maximum current level that can be drawn at this interval. The estimated level per thread is sent (with the

VEL instruction) to the PMU as shown in Fig. 4.10, and the PMU handles the alignment cases by setting the appropriate voltage level based on number of cores having high-power event. The voltage guardband is a function of the number of cores sharing the same VR that reports high voltage emergency level. This is due to the fact that at a given time interval, the total current that is consumed from a shard VR between N cores equals to the sum of current consumption by each core. For example, if one core has high voltage emergency level (VEL), then the PMU adds additional 10 mV voltage guardband to the nominal voltage, while if there are three cores that report high voltage emergency level, then the PMU adds 30 mV voltage guardband to the nominal voltage. For doing the guardband calculation, the PMU needs to know the PDN topology of the processor, how many cores the systems have, and which cores share the same VR or have separate VR.

In simultaneous multithreading (SMT), instructions from more than one thread can be executed in any given pipeline stage at a time. At SMT case, each software thread will set the VEL (a value between 0 and 1 based on running code-estimated energy), the PMU sums the VEL value from both threads and determines if a voltage emergency is expected based on the threshold. Although the developed method takes multithreading and multi-core into account, we focus on single-thread workloads in this work and keep multithreading experiments for future work.

4.8 Conclusions

Power delivery is a significant constraint for high-performance CPUs. Building the power delivery network for the worst-case power excursion is costly and energy inefficient. We developed a novel compiler-assisted method that identifies code intervals with high-power excursion risk. This method runs on a standard CPU design with minor addition to existing power management techniques and does not require special design techniques or significant architectural changes as proposed by previous works. It is limited to identifying power excursion risks. These power excursions may or may not materialize at runtime, depending on actual data, control flow, and cache misses. We have implemented the method through LLVM compiler and have tested it on a 4th Generation Intel® Core™ processor. We have demonstrated performance gain of up to 16% for a power delivery constrained platform and up to 11.4% energy savings for a power delivery capable platform. We have validated the implementation safety and found no unidentified power excursions. The fundamental limitation of compile-time technique was insignificant, falsely marking only 1.07% of the low-power intervals as having high-power risk.

References

1. Yahalom, Gilad, Omer Vikinski, and Gregory Sizikov. Architecture constraints over dynamic current consumption. Electrical Performance of Electronic Packaging, 2008 IEEE-EPEP. IEEE, 2008.
2. Wechsler, O. 2006. Inside Intel® Core™ microarchitecture: Setting new standards for energy-efficient performance. *Technology*, 1.
3. Charles, J., Jassi, P., Ananth, N. S., Sadat, A., & Fedorova, A. 2009, October. Evaluation of the Intel® Core™ i7 Turbo Boost feature. In *Workload Characterization, 2009. IISWC 2009. IEEE International Symposium on* (pp. 188–197). IEEE.
4. Lattner, Chris, and Vikram Adve. "LLVM: A compilation framework for lifelong program analysis & transformation." Code Generation and Optimization, 2004. CGO 2004. International Symposium on. IEEE, 2004.
5. Jain, Tarush, and Tanmay Agrawal. "The Haswell Microarchitecture-4th Generation Processo".
6. Hammarlund, Per, et al. "4th Generation Intel® Core™ Processor, codenamed Haswell." (2013): 1–1.
7. Jawad Haj-Yihia, Power Profiling of third droop voltage-emergencies tool, 2014, from Haifa University: https://drive.google.com/folderview?id=0B3IgzCqRS5Q_NDZ0dWxZeTdHV2c&usp=sharing.
8. Brooks and M. Martonosi, "Dynamic thermal management for high-performance microprocessors," in Proc. Int. Symp. On High Performance Computer Architecture, Jan. 2001, pp. 171–182.
9. K. Skadron, "Hybrid architectural dynamic thermal management," in Proc. Design, Automation & Test in Europe Conf., Mar. 2004.
10. S. Heo, K. Barr, and K. Asanovic, "Reducing power density through activity migration," in Proc. Int. Symp. Low Power Electronics & Design, Aug. 2003, pp. 217–222.
11. Intel while paper, "Measuring Processor Power, TDP vs. ACP" http://www.intel.com/content/dam/doc/white-paper/resources-xeon-measuring-processor-power-paper.pdf, 2011.
12. M. Popovich, A. V. Mezhiba, and E. G. Friedman, Power Distribution Networks With On-Chip Decoupling Capacitors. New York: Springer, 2008.
13. Patrik Larsson, "Resonance and Damping in CMOS Circuits with On-Chip Decoupling Capacitance," IEEE Trans. on CAS-I, pp. 849–858, Aug. 1998.
14. Reddi, Vijay Janapa, and Meeta Sharma Gupta. "Resilient Architecture Design for Voltage Variation." Synthesis Lectures on Computer Architecture 8.2 (2013): 1–138.
15. Kim, Youngtaek, et al. "Audit: Stress testing the automatic way." Microarchitecture (MICRO), 2012 45th Annual IEEE/ACM International Symposium on. IEEE, 2012.
16. Intel Corporation. 2009. Voltage Regulator Module (VRM) and Enterprise Voltage Regulator-Down (EVRD) 11.1 Design Guidelines. Reference Number 321736, Revision 002, September, 2009.
17. Zhang, Michael T. "Powering Intel (r) Pentium (r) 4 generation processors." Electrical Performance of Electronic Packaging, 2001. IEEE, 2001.
18. Miller, Timothy N., et al. "VRSync: characterizing and eliminating synchronization-induced voltage emergencies in many-core processors." ACM SIGARCH Computer Architecture News. Vol. 40. No. 3. IEEE Computer Society, 2012.
19. Kanev, Svilen, et al. "Measuring code optimization impact on voltage noise." Change 40 (2013): 20.
20. Reddi, Vijay Janapa, et al. "Predicting voltage droops using recurring program and microarchitectural event activity." IEEE micro 30.1 (2010): 110.

21. Lefurgy, Charles R., et al. "Active management of timing guardband to save energy in POWER7." proceedings of the 44th Annual IEEE/ACM International Symposium on Microarchitecture. ACM, 2011.
22. Austin, Todd M. "DIVA: A reliable substrate for deep submicron microarchitecture design." Microarchitecture, 1999. MICRO-32. Proceedings. 32nd Annual International Symposium on. IEEE, 1999.
23. Mukherjee, Shubhendu S., Michael Kontz, and Steven K. Reinhardt. "Detailed design and evaluation of redundant multi-threading alternatives." Computer Architecture, 2002. Proceedings. 29th Annual International Symposium on. IEEE, 2002.
24. Reddi, Vijay Janapa, et al. "Voltage emergency prediction: Using signatures to reduce operating margins." High Performance Computer Architecture, 2009. HPCA 2009. IEEE 15th International Symposium on. IEEE, 2009.
25. Joseph, Russ, David Brooks, and Margaret Martonosi. "Control techniques to eliminate voltage emergencies in high performance processors." High-Performance Computer Architecture, 2003. HPCA-9 2003. Proceedings. The Ninth International Symposium on. IEEE, 2003.
26. Toburen, Mark C. "Power analysis and instruction scheduling for reduced di/dt in the execution core of high-performance microprocessors." 1999.
27. SPEC 2006; Standard Performance Evaluation Corporation, www.spec.org/.
28. Firasta, Nadeem, et al. "Intel avx: New frontiers in performance improvements and energy efficiency." *Intel white paper*, 2008.
29. Shao, Yakun Sophia, and David Brooks. "Energy characterization and instruction-level energy model of Intel's Xeon Phi processor." Low Power Electronics and Design (ISLPED), 2013 IEEE International Symposium on. IEEE, 2013.
30. Hähnel, Marcus, et al. "Measuring energy consumption for short code paths using RAPL." ACM SIGMETRICS Performance Evaluation Review 40.3 (2012): 13–17.
31. A. Yasin, "A Top-Down Method for Performance Analysis and Counters Architecture," presented at the Performance Analysis of Systems and Software (ISPASS), 2014 IEEE International Symposium on, 2014.
32. Kihwan C., Soma, R., Pedram, M. Fine-grained dynamic voltage and frequency scaling for precise energy and performance tradeoff based on the ratio of off-chip access to on-chip computation times. Computer-Aided Design of Integrated Circuits and Systems, IEEE Transactions on, 24(1), 18–28, January 2005.
33. Grochowski, Ed, Dave Ayers, and Vivek Tiwari. "Microarchitectural simulation and control of di/dt-induced power supply voltage variation." High-Performance Computer Architecture, 2002. Proceedings. Eighth International Symposium on. IEEE, 2002.
34. Gupta, Meeta Sharma, et al. "DeCoR: A delayed commit and rollback mechanism for handling inductive noise in processors." High Performance Computer Architecture, 2008. HPCA 2008. IEEE 14th International Symposium on. IEEE, 2008.
35. Ernst, Dan, et al. "Razor: A low-power pipeline based on circuit-level timing speculation." Microarchitecture, 2003. MICRO-36. Proceedings. 36th Annual IEEE/ACM International Symposium on. IEEE, 2003.
36. Ernst, Dan, et al. "Razor: circuit-level correction of timing errors for low-power operation." IEEE Micro 24.6 (2004): 10–20.
37. Gupta, Meeta Sharma, et al. "Towards a software approach to mitigate voltage emergencies." Low Power Electronics and Design (ISLPED), 2007 ACM/IEEE International Symposium on. IEEE, 2007.
38. Reddi, Vijay Janapa, et al. "Eliminating voltage emergencies via software-guided code transformations." ACM Transactions on Architecture and Code Optimization (TACO) 7.2 (2010b): 12.

39. Reddi, Vijay Janapa, et al. "Voltage smoothing: Characterizing and mitigating voltage noise in production processors via software-guided thread scheduling." Microarchitecture (MICRO), 2010 43rd Annual IEEE/ACM International Symposium on. IEEE, 2010a.
40. Reddi, Vijay Janapa, et al. "Voltage noise in production processors." IEEE micro 31.1 (2011): 20–28.
41. Intel 64 and IA-32 Architectures Software Developer's Manual, Volume 3, Section 14.9 as of August 2014.
42. Kim, Youngtaek. Characterization and management of voltage noise in multi-core, multi-threaded processors. Diss. 2013.

Chapter 5
Static Power Modeling for Modern Processor

5.1 Introduction

Energy efficiency has become one of the most important design parameters for hardware and software developers, due to battery life on mobile devices and energy costs and power provisioning in data centers. Energy and power estimation of software are essential tools for optimizing programs running on both general-purpose and embedded computer systems. These tools enable systems designers to choose the correct design parameters for their systems, such as voltage regulators, board design, and cooling solutions. In addition, these tools help software developers to find the code areas with high power consumption and to optimize their code based on feedback from the tool.

Static power and energy estimation tools can be used at compile-time to give the compiler hints on which compiler flags and optimizations can minimize the power or energy of the compiled program.

In addition, these tools allow software developers to find "power/energy bugs" at their programs by observing power consumption of the software at function level and loop level over time. This helps with understanding the power behavior of the developer's program, and hint on possible optimizations. For example, the tool can point to loops that generate high cache misses and give, hints of enabling or disabling the hardware prefetching before entering these loops [7]. Other example is the ability of these tools to point on paths or sub-paths that have the highest power, this can be used for voltage smoothing and voltage emergency prediction [5, 6].

Another area of optimization is the dynamic optimization. Compiler or optimization tools can instrument code to change the voltage and frequency of the processor in order to improve the energy efficiency [4, 8, 9].

One of the disadvantages of the current energy estimation tools is that they work on one path at a time for the running program. The path depends on the input of the program, therefore, in order to cover the whole program paths, the user needs to create stimulus and verify full paths coverage. This is also the case for the memory

© Springer Nature Singapore Pte Ltd. 2018
J. Haj-Yahya et al., *Energy Efficient High Performance Processors*,
Computer Architecture and Design Methodologies,
https://doi.org/10.1007/978-981-10-8554-3_5

accesses, in some runs, a memory access might cause a cache miss while in other runs with other input the memory access will result in a cache hit. For this reason, a method that checks all the possible program paths and inputs will give more accurate analysis and wider observation of the whole program behavior.

The work proposed in [19, 25] discusses energy-aware programming support via symbolic execution. For each code path explored by symbolic execution tool, the base energy cost can be highlighted to the programmer. The energy consumption is based on vendor profiles provided by Android device manufacturers. However, such an approach is preliminary in the sense that it only considers the core power consumption. In contrast, power consumption to access memory subsystems, network card and other I/O components were not considered. In this work, we have modeled the core and cache power consumption and showed how the I/O and the other components of the processor can be modeled.

Through this work, we have developed a Symbolic Execution tool for Energy Modeling (SEEM) for High-Performance Computing (HPC) processor. The tool enables power and energy-aware compilation. It gives statistics on the core and cache power and energy at the various paths of the program, which enables compiling the program with different compilation flags, and then choosing the compilation parameters that minimize the energy, power or other parameters that combine power, and energy and performance. The SEEM system can execute programs symbolically. Instead of using concrete values, SEEM uses symbolic and concrete values and converts instructions into constraints that are being solved with constraint-solving techniques (SAT [10]).

The SEEM system uses an energy model, which includes energy modeling of the core and cache subsystems. This enables the tool to report energy and power statistic of the various program paths and can potentially give hints to the user on possible optimizations. The main challenges with static energy estimation of programs are the mostly unknown number of iterations and the result of conditional branches caused by unknown input data. This means the program flow varies and consequentially, the time consumption of the software component in total. Large estimation errors are possible because of this. Another challenge is to estimate the energy consumption of the cache, this energy is usually a function of cache miss and hit of the memory accesses due to the microarchitecture implementation of the memory subsystem.

This work outlines the following contributions:

- We have developed a SEEM, this includes building an instruction-level energy model for the 3rd Generation Intel® Core™ i7 3517U Processor code name IvyBridge. The model predicts energy consumption of core and cache with an error rate of up to 13%. To the best of our knowledge, an energy model that uses symbolic execution for high-performance CPUs has not been done before.
- We characterize the Energy Per Instruction (EPI) of IvyBridge processor using a set of specialized microbenchmarks exercising different categories of instructions.

- The tool gives statistics on the energy and power consumption of the program's feasible paths for the core and cache subsystem, for example, it gives the maximum path-power among all feasible paths.

This tool provides system developers opportunities to improve energy efficiency. In particular, the tool gives statistics on the energy consumption of the program's paths. The tool enables compiling the program with different compilation parameters and choose the one that minimizes the energy or power.

5.2 Symbolic Execution

To estimate the energy and power consumption of a program we use a symbolic execution method. For each instruction, we associated an EPI value and we evaluate the energy and power that each path consumes. The energy model for the instructions is similar to Tiwari's model [2] and Neville's [14] model. The proposed energy model was applied at the Intermediate Representation (IR) instructions of the LLVM [15] compiler (LLVM IR) rather than at instruction set of specific processor's architecture. The effect of the switching cost between instructions is approximated into the actual instruction cost, rather than assigning a unique overhead for each instruction pairing.

5.2.1 Why Symbolic Execution?

Static analysis—symbolically executing a program can be used to statically analyze the program energy and power without running a real system, this method can be applied at compilation stage (e.g. checking the effect of different compilation flags on energy and power).

   Cover program's feasible paths—Program execution paths vary with different inputs to the program, for covering all possible paths, we need to run many times with different inputs to have reasonable coverage of program paths, different programs paths have different energy and power values. Using symbolic execution we will be able to get statistics about the whole program, for example, the maximum power that can be drawn by some path or sub-path, this statistic can be used for power capping [18] and voltage emergency prediction [9]. Running the program symbolically more accurate than analyzing the control flow graph (CFG) of the program as symbolic execution traverse feasible paths and not all possible paths at the CFG. Much research [34] was done on flow analysis algorithms to detect infeasible program paths.

   Solving symbolic queries—the symbolic execution system includes SAT solver [10]). The SAT solver enabling solving queries related to path condition and to memory addresses. The relationship between memory addresses can be predicted in case of concrete or symbolic addresses. Cache hit and miss can be predicted by monitoring memory accesses instruction (load and store) and symbolically model the cache behavior.

5.2.2 How Symbolic Execution Works?

Symbolic execution [3] is a path-sensitive program analysis technique, it computes program output values as expressions over symbolic input values and constants.

The state of a symbolic execution is a triple (l, PC, s), where:

- l is the current location, records the next statement to be executed,
- PC is the path condition (Do not confuse with program counter), is the conjunction of branch conditions encoded as constraints along the current execution path
- s:M × expr is a map that records a symbolic expression for each memory location, M, accessed along the path.

For computation statements, $m1 = m_2 \oplus m_3$, where the $m_i \in M$ and $\oplus$ is some operator, when executed symbolically in state (l, PC, s) produce a new state (l + 1, PC, s'), where $\forall m \in M - \{m_1\}$: s'(m) = s(m) and $s(m_1) = s(m_2) \oplus s(m_3)$, i.e., state not changed for none m_1 memory locations and m_1 state is affected by the operation.

For Branching statements, if $m_1 \oplus m_2$ goto d, when executed symbolically in state (l, PC, s) branch the symbolic execution to two new states (d; PC ∧ (s $(m_1) \oplus s(m_2)$); s) and (l + 1; PC ∧ ¬ $(s(m_1) \oplus s(m_2))$, s) corresponding to the "true" and "false" evaluation of the branch condition, respectively.

An automated decision procedure (e.g. SAT solver [10]) is used to check the satiability of the updated path conditions, when a path condition is found to be unsatisfiable; symbolic execution along that path halts.

Symbolic execution of the code fragment in Fig. 5.1 uses symbolic values X and Y to denote the value of variable x and y, respectively on entry to the code fragment. Symbolic execution determines that there are two possible paths: the first path is when X > Y and the second path is when X ≤ Y. At the right side of Fig. 5.1 we can see the execution path and the path condition (PC) after executing each instruction. When the first path (X > Y) arrive to the second "if" statement, the values of x and y variables are Y and X symbolic values, respectively. At this point, the path forks to two paths, where each path has updated path condition. Solving the path condition predicate for the assert statement results in False, meaning that this path is not feasible. Symbolic execution in this example is used to check code correctness at all possible input values.

While symbolic execution is powerful program analysis technique, it has a fundamental restriction. The main challenge with symbolic execution is the path explosion. Symbolically executing all feasible program paths does not scale well

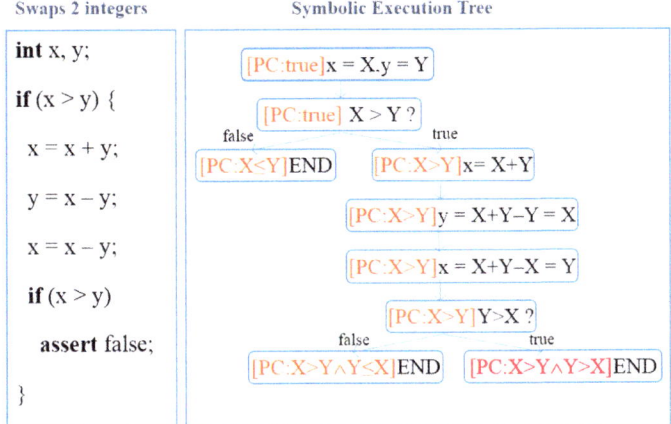

Fig. 5.1 Example of symbolic execution of a code

with large programs. The number of feasible paths in a program grows exponentially with an increase in program size and can even be infinite in the case of programs with unbounded loop iterations [11]. Figure 5.2 shows an example of symbolic execution with unbounded loop, the loop bound (n) is symbol (with symbolic value S), in this case, the number of execution paths is infinite. Solutions to the path explosion problem generally use either heuristics for pathfinding to increase code coverage [12], bounding execution time of a single path, bounding the range of symbolic values or reduce execution time by parallelizing independent paths [13].

Fig. 5.2 Symbolic execution with loops

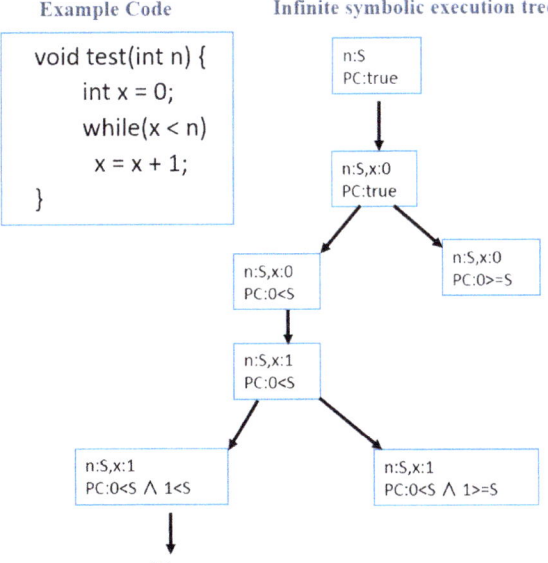

5.3 Problem Statement

Given a program, the goal is to report statistics on the power and energy of the program without running the program on real system (static analysis), these statistics includes: maximum power/energy, average power/energy, power/energy sum over all feasible paths of the program, and other user-defined statistics related to power or energy; such statistics can be used, for example, at compilation stage to choose compilation flags that minimize one of the metrics; moreover it can be used by system designers to obtain statistics about the maximum power that the program can draw, which will help them to take decisions on cooling solutions and power delivery options.

5.3.1 Definitions

Given a directed graph G (possibly with cycles) with the control flow graph such that:

- G has a start-node s (program entry) and end node e (program exit).
- Every node v has a weight that is represented by the function EPI(v) which denotes Energy Per Instruction.
- Let $P = v_1 \rightarrow v_2 \rightarrow \ldots \rightarrow v_n$ be a path of size n possibly with cycles, where V_1 is the start-node (s) and V_n is the end node (e).

We define the following:
Path-Energy (E_{path}):

$$E_{path} = \sum_{i=1}^{n} EPI(V_i)$$

Path-Power[1] (P_{path}):

$$P_{path} = \sum_{i=1}^{n} \frac{EPI(V_i)}{n},$$

---

[1]In this work, we assume that the cycles per LLVM instruction (C) are constant for all instructions.

where:

$$EPI(Instruction) = \begin{cases} Mul, & mulEnergy \\ Add, & addEnergy \\ \ldots; & \ldots \\ Div, & divEnergy \\ mem(hit), & hitEnergy \\ mem(miss), & missEnergy \end{cases}$$

Other definitions of path-energy and path-power are possible per user need, e.g., path-power can be defined as the maximum power for every consecutive K instruction at the path, this metric can be used to observe the peak power that the program can draw at given interval of time (K instructions).

5.3.2 Programs Infeasible Paths

Each execution of the program follows one path through its control flow graph (CFG). In general, a program has a large number of such paths. Regardless of the quality of the program and the programming language used to develop it, in general, a sizable number of these paths are infeasible—that no input can exercise them [34]. At Fig. 5.3 we have two real programs codes, the first code (A) is a code that implementing a function implements absolute value calculation. We can see that the

```
   //-- Code A ----------
1.   int abs(int a) {
2.      int abs_val;
3.      if(a>=0)
4.         abs_val=a;
5.      if(a<0)
6.         abs_val=-a;
7.      return abs_val;
8.   }

   //-- Code B ----------
1.   If(sameGoto)
2.      newTarget = ((IfStmt) stmtSeq[5]).getTarget();
3.   else {
4.      newTarget = next;
5.      oldTarget = ((IfStmt) stmtSeq[5]).getTarget();
6.   }
7.   ...
8.   If(IsameGoto)
9.      b.getUnits().insertAfter(... );
10. ...
```

Fig. 5.3 Infeasible paths examples

path $1 \rightarrow 2 \rightarrow 3 \rightarrow 4 \rightarrow 5 \rightarrow 6 \rightarrow 7$ is infeasible. The second code(B) is taken from Soot [35] application, we can see that the path $1 \rightarrow 2 \rightarrow 7 \rightarrow 8 \rightarrow 9$ is infeasible.

5.3.3 Dataflow Versus Symbolic Execution Analysis

Analyzing the dataflow graph of a program can give us visibility on programs paths, but the limitation with this method is that it traverses feasible and non-feasible paths, while the symbolic execution traverses feasible paths only.

In Fig. 5.4 we have a program that reads a value into variable N and executes vector or scalar instructions depending on the value of N. Near each instruction, we have the EPI and in the circle, we have instruction number. Dataflow graph analysis would give us four paths $(1 \rightarrow 2 \rightarrow 3 \rightarrow 5 \rightarrow 6, \quad 1 \rightarrow 2 \rightarrow 3 \rightarrow 5 \rightarrow 7, \quad 1 \rightarrow 2 \rightarrow 4 \rightarrow 5 \rightarrow 6, \quad 1 \rightarrow 2 \rightarrow 4 \rightarrow 5 \rightarrow 7)$ and the energy for these paths based on our previous definition is (23, 18, 18, 13), respectively, so the maximum energy per all paths is 23. While with deeper analysis, we can see that the path

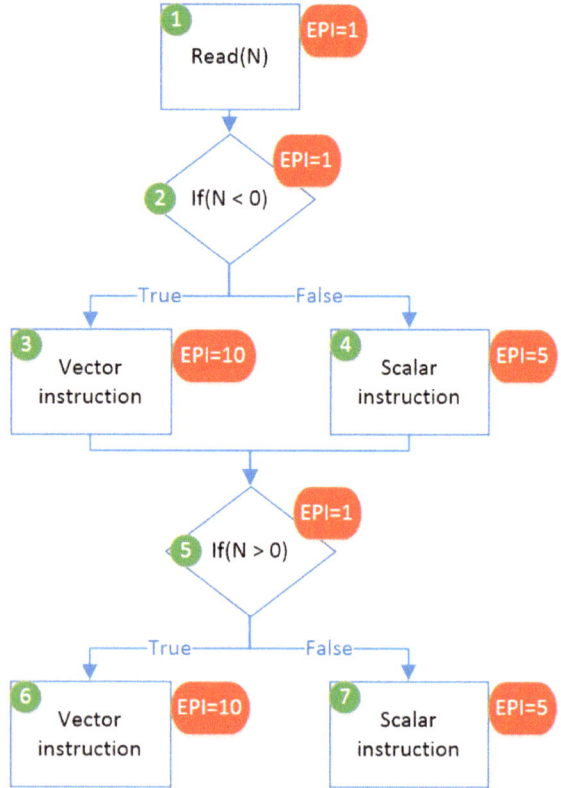

Fig. 5.4 Symbolic execution versus dataflow analysis

$(1 \rightarrow 2 \rightarrow 3 \rightarrow 5 \rightarrow 6)$ is infeasible and running the program with symbolic execution analysis will eliminate this infeasible path, hence the correct maximum energy per all paths is 18 and not 23. This example demonstrated how symbolic execution analysis is more accurate than dataflow analysis.

5.3.4   Symbolic Execution and Power/Energy Statistics Example

In this section, we give a detailed example of the method that we used to estimate the power and energy of the core and cache using symbolic execution.

In Fig. 5.5 we have a code snippet that under certain conditions does operations using two arrays (A, B) and put the result into the first array (A). The code reads two numbers (user input), the first number is used as a limit for the first loop and the second number is used as an offset of array B.

Figure 5.6 shows the symbolic execution tree of the code snippet. The variables n and s are defined as symbolic variables and they get symbolic values of N and S, respectively. The memory state appears near each operation, and right to it appears the path-condition (appears when changed).

In this example, we **bounded N to be between 0 and 3** (including) and the array "A" is assumed to have the values {1, 2, 3} at the beginning of the code. The execution tree at Fig. 5.6 has four paths, each path starts from the root and finishes at one of the four rectangles labeled with "END".

To calculate the power and energy of each path, we use the previously defined E_{path} and P_{path} equations; we differentiate between core and cache domains statistics, such that each domain has different EPI; for example, the "add" instruction between registers consumes power at the core-domain while does not consume power at the cache domain, hence, the $EPI_{core}(add)$ is positive number while $EPI_{cache}(add)$ is nearly zero[2]; on the other hand, for the "load" instruction the EPI will be positive for both core and cache. For simplicity, we assume that

- EPI_{core} for addition instructions (+/++) is 5 μJ and for the rest of the instructions, it is 1 μJ.
- EPI_{cache} for load/store with cache Hit/Miss is 1/10 μJ, respectively and for the rest of the instructions, it is 0.

Based on the above assumptions, the **core** energy and power of the shortest and longest paths are as follows:

- Shortest path core energy is 4 μJ:
  - four instructions with EPI = 1 for each instruction.

---

[2]There might be leakage or dynamic power that is consumed by the cache regardless of the executed instruction at the core, this depends on the cache power management implementation.

```
read(n);
read(s);
for(i=0;i<n;i++)
  if(i%2 == 0)
    for(j=0;j<=i;j++)
      A[j]+=B[s+j];
```

Fig. 5.5 Example code to illustrate our method

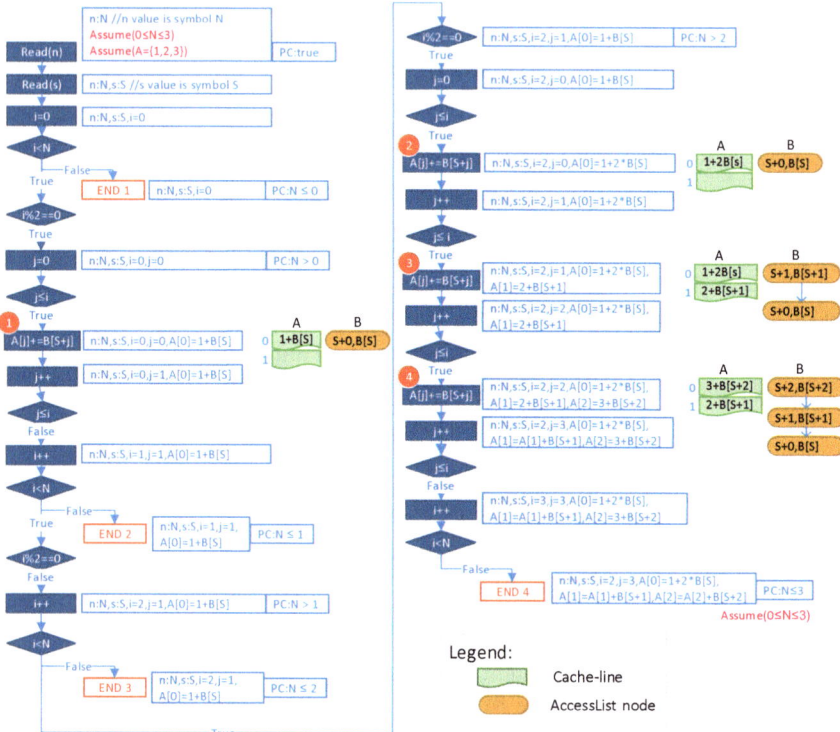

Fig. 5.6 Example of power and energy estimation using symbolic execution

- Shortest path core power is 1 μW:

  - four instructions with EPI = 1 for each instruction divided by the number of instructions.

- Longest path core energy is 85 μJ

  - Here, we counted each A[j] += B[s + j] as four instructions: (two loads, store, and addition), we have 11 addition instructions and 30 none addition.

- Longest path core power is 2.1 μW

  - 85 divided by 41 instructions.

To calculate the energy and power of the **cache**, we need to predict if the memory access (load/store) has a cache hit or miss because the EPI is different for each case. To predict access status, we emulate cache behavior, at this example, we emulate simple cache with two entries to illustrate our method. For concrete memory accesses (load/store with concrete array index access); as we have at access A[j]; we use regular cache per each array, and we use the array index as the address to access the cache (treat the index as Tag:Set:Offset based on cache parameters). For example, access to A[0] will target first entry (cache Set 0) and access A [1] will target second entry (cache Set 1) at our two entries cache. For symbolic memory access, as we have at access B[s + j], we use access list, which records in ordered list all symbolic accesses to a given array (array B in our example) as a pair of index-expression and value expression (index-exp, value-exp). In Sect. 5.4.2 (cache emulator) we show the algorithm to predict if a given access was a cache hit or miss, at the following example, we will illustrate our method.

5.3.4.1 Concrete Memory Access

For memory access with **concrete index**, the cache of the array will be accessed. At our example, all accesses to array A are concrete accesses, in Fig. 5.6 the cache of A appears right to memory instructions. The prediction for the longest path access in our example is as follows:

- At array access region labeled with 1 (inside circle) we have load from A[0] and store to A[0], the first access is miss, A[0] will be entered to the cache at Set 0, and the second access (store) is hit.
- The accesses labeled by 2 also have one load and store to A[0] and both accesses are hit (as A[0] is in the cache).
- The accesses labeled by 3 is to A [1] (load followed by store), the load is miss, A [1] will be entered to the cache at Set 1 and the store is hit.
- The accesses labeled by 4 is to A [2], the load will be miss, A [2] will replace A [0] at the cache and will be entered to Set 0, the store will be hit.

In total for the concrete memory access we have five hits and three misses.

5.3.4.2 Symbolic Memory Access

For memory access with **symbolic index**, the AccessList of the memory location will be accessed. AccessList is a data structure that consists of ordered list of the memory accesses, each node at the list is a pair (index, value), where the index is symbolic expression and the value is a symbolic expression or concrete value. The AccessList will be traversed in order to answer two questions

1. Was the cache line corresponding to this access evicted from cache?
2. Does the accessed location exist in the cache?

For memory access with symbolic index (*sym-index*), the questions above can be answered by checking the AccessList. Traverse the AccessList from the newest access to the oldest access, and for each node (*index-expr, value-expr*)

- Create a predicate that compares the *Set* field of index-expr with the *Set* field of *sym-index*, if this predicate is satisfiable (by SAT solver) then, we check the *tag* fields by creating additional predicate that compares the *Tag* field of *index-expr* and *Tag* field of sym-index, if the Tag predicate is satisfiable then, we predict cache hit, and if the Tag doesnot match then, this access had evicted the cache line that includes our address,[3] therefore, we predict cache miss (and we stop the AccessList traverse).
- If no hit was found at the AccessList traverse then we predict cache miss for this memory access.

In our example, all accesses to array B are symbolic accesses. In Fig. 5.6 the AccessList of array B appears right to memory instructions. The memory access prediction for the **longest path** is as follows:

- The accesses labeled by 1 is to B[S] (load), the AccessList of B is empty, so we predict miss for this access and enter a pair of (S, B[S]) into the AccessList as a pair of index-expression and value-expression.
- The accesses labeled by 2 is to B[S], we traverse the AccessList and compare the *Set* field of S expression to *Set* field at each index-expression at the AccessList (Set field is bit 0 of the address because we are emulating cache with two entries), a predicate that compares bit 0 of S with bit 0 of S (first node at AccessList) is satisfiable, then, the Tag bits (1–31) are compared and this predicate is also satisfiable, therefore, we predict hit and we enter the pair (S, B[S]) into the AccessList.
- The accesses labeled by 3 is to B[S + 1], we traverse the AccessList and compare the Set field of S + 1 expression to Set field at each index-expression at the list (Set field is bit 0 of the address because we are emulating cache with 2 entries) hence when comparing bit 0 of (S + 1) with bit 0 of (S) then it doesn't match (predicate is not satisfiable). So we predict miss and we enter the pair (S + 1, B[S + 1]) into the AccessList.
- The accesses labeled by 4 is to B[S + 2], we traverse the AccessList and compare the Set field of S + 2 expression to Set field at each index-expression at the list, in our case the Set field is bit 0 of the address (because we are emulating cache with two entries) hence, when comparing bit 0 of (S + 2) with bit 0 of

---

[3]More details in cache emulator section.

(S + 1) then, it not satisfiable. When comparing next node (with index-expression S) bit 0 of (S + 2) with bit 0 of S the predicate is satisfiable, so we compare the Tag fields (bits 1–31) of both addresses, this predicate is not satisfiable, so we declare miss and we enter the pair (S + 2, B[S + 2]) into the AccessList.

- In case that the code access after that B[S] (does not appear the code or at Fig. 5.6) then, although the index S exist at the AccessList, this access will miss the cache, because at first stage we will compare the access index (S) to the nodes of AccessList and the first node to be compared is S + 2, comparing the *Set* fields (bit 0) of the two address matches but the *Tag* fields will not match, this means that we predict that the cache line was evicted and we stop the search and declare miss.

In total, for the symbolic memory access we have one hit and three misses.

In total for the combined memory access (concrete and symbolic) we have six hits and six misses for the 12 memory accesses.

Based on the above analysis, the cache energy and power of the longest paths (with 41 instructions) are as follows:

- Longest path's Cache energy is 66 µJ:

  - six hits with $EPI_{hit} = 1$ and six misses with $EPI_{miss} = 10$.

- Longest path's cache power is 1.6 µW:

  - 66 divided by the number of instructions on the path (41 instructions).

5.3.5   Memory Accesses and SAT Solver

Symbolic execution is enabled due to the ability of modern powerful SAT solver to solve complicated predicates. During the symbolic execution, we are trying to determine if a certain formula is satisfiable, for example, is a particular program point reachable? This can be determined by checking with SAT solver of the path condition (PC) is satisfiable. Other example is checking if array access A[i] is out of bound or not, this can be figured by conjunction of the path condition and i < 0 ∨ i > A_length (length of the array A), this predicate is passed to SAT solver, if this predicate is satisfiable then, there is a case where the array will be accessed out of bound.

In addition to the above usages, we use the power of SAT solver to solve queries on memory accesses in order to determine if memory access will have cache hit or cache miss as shown in the previous example. Predicates on addresses might be more complicated and hard to solve them without SAT solver.

For example, if one of the previous accesses to an array A was done with symbolic index I (i.e. A[I]) and next access was done to index f(I) (i.e. A[f(I)]), in order to determine if the cache line for the address A + f(I) exists already on the cache, we need to check if the T\]ag and Set fields of the two **symbolic** address f(I) and I matches, in order to do that we need to solve a predicate on these symbolic addresses with SAT solver, assuming that the Set and Tag fields are bits 15:7 and 31:16, respectively, then, we need to solve the following two predicates in order to determine if the particular cache line exists in the cache

1. I[15:7] == f(I)[15:7]
2. I[31:16] == f(I)[31:16]

If the first predicate is satisfiable but the second is not, this means that the cache line was evicted and the access f(I) is marked as miss because a new cache line with the same Set was accessed, while if the two predicates are satisfiable this means that the cache line exists in the cache and the access f(I) is marked as hit. Solving the two above predicates requires SAT solver as the function f might not be simple function.

5.4 Framework

The SEEM framework consists of LLVM compiler, energy-model, cache emulator and symbolic execution engine as shown in Fig. 5.7.

5.4.1 Energy Model and Power Simulator

The energy model that we have built is processor specific, which means that moving to another processor will require the recalibration of the model. In this work, we are targeting the 3rd Generation Intel® Core™ Processor code name IvyBridge.

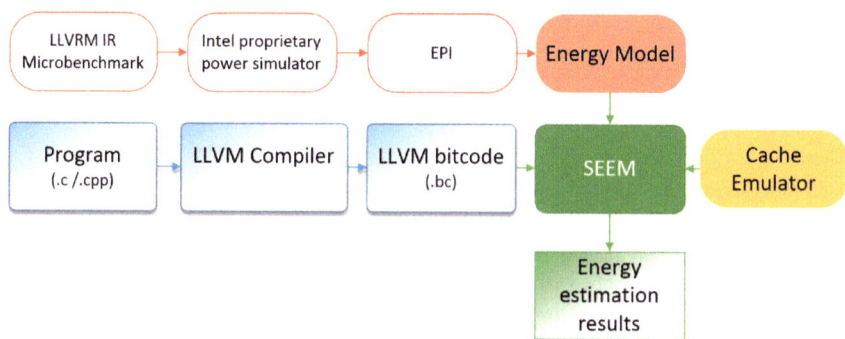

Fig. 5.7 SEEM framework for power estimation using symbolic execution

To create the energy model we used a generic method that was used by other works [2, 16]. The method builds energy model based on microbenchmarks. Each microbenchmark is a loop that iterates a target instruction type. Another type of microbenchmarks was developed to find the energy of memory access with cache miss or hit. The benchmark allocates array that is multiple cache line size. For cache hit benchmark, the loop traverses part of the array (up to cache-line size), consecutively, this will generate one cache miss while all other access hit the cache. For measuring the energy of cache miss the array is traversed with strides of cache line in order to generate cache miss every memory access. The energy of the running microbenchmark is being measured and divided by the number of the loop iterations to find the EPI of a single instruction.

In our work, in order to find the energy of the core and the level 1 cache (L1) separately for the running microbenchmarks, we used an Intel proprietary power simulator. The power simulator is a microarchitectural simulator, and it uses low-level design data and microarchitectural events to estimate the power of the assembly code running. The Intel power simulator has accuracy of 3–5% for performance workloads (i.e. workloads without CPU idleness). Unlike the RAPL [21] energy reporting counters that reports combined energy number for all the cores and Last Level Cache (LLC), the Intel proprietary power simulator reports power breakdown for the core, L1, and other core subdomains, this data is essential for our model generation. Typically, this simulator is used to run microbenchmark and not large programs. We ran the microbenchmarks on the simulator and used the energy data outputted by the simulator to find the EPI for each LLVM IR instruction and for cache miss and hit for memory instructions.

5.4.2 Cache Emulator

SEEM extended the symbolic execution to estimate the cache energy and power. For this purpose, we have developed a cache emulator within our tool. The cache emulator imitates the behavior of the first-level cache (L1) in order to predict if a memory access will have a hit or miss at the cache. The prediction then propagates to the energy estimator that adds the appropriate EPI to the energy calculation as described in Sect. 5.3.

5.4.2.1 Memory and Cache Structure

Memory model is a mapping between memory *allocation-site* and *memory-values*:

- *allocation-sites* are variables or dynamic memory allocations (calls to malloc, stack objects, and global variables), conceptually, can be thought of as the unique name for the object allocated at that site. *allocation-sites* can be accessed with indexes, index represents the offset from the address at which *allocation-sites* resides at memory. For example, array[i]—where array is the *allocation-site*s and i is the index. Additional example is $P = P \rightarrow Next$, in this case, the $P \rightarrow Next$ is represented as allocation-site P and index representing the offset of struct field *Next*. Index for allocation-site can be concrete or symbolic.

- **memory-values** consist of a data structure that holds the memory access information that is related to the *allocation-site* (e.g. array and all accesses to its indices). *The memory-values* can be either concrete or symbolic.

Each *allocation-site* is given a separate mapping in memory.
Memory accesses (load/store) can be represented as follows:

- load allocation-site, index
- store allocation-site, index, and memory-value

Memory accesses (load/store) with concrete index target one-way virtually-addressed cache as shown in Fig. 5.8. While in case of memory access with symbolic index, the access is recorded at *AccessList* data structure as shown in Fig. 5.8.

The *AccessList* data structure is an ordered list of the memory accesses that is maintained for each memory *allocation-site* in the program, each node at the list is a pair (index, value), where the index is a symbolic expression and the value is a symbolic expression or concrete value of the memory location content.

5.4.2.2 Memory Access

For concrete accesses and for symbolic accesses at concrete indexes, a simple, virtually-addressed cache is being accessed, according to the Tag, Set, and Offset of accessed address as shown in Fig. 5.8.

Once the programs try to access memory with symbolic address (e.g. array with symbolic index) then the address is being compared to the corresponding *AccessList* indexes, and an expression is being calculated and returned, and a new node is being added to the head of the list to record the memory access.

5.4.2.3 Cache Hit or Miss

As shown in Fig. 5.8, once a memory access (load or store) instruction is executed, we first locate the allocation-site for this access, then we use the index part (denoted by INDX in Fig. 5.8) of the memory access to access the *memory-value's* data structure, we treat the INDX as address that break into Tag, Set, and Offset in order to access the cache. We differentiate between concrete and symbolic addresses as follows:

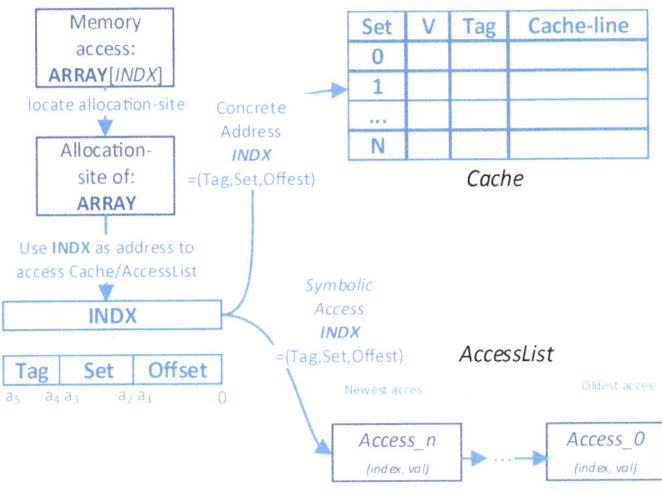

Fig. 5.8 Cache modeling for concrete and symbolic accesses

1. In case that the address is concrete then we check if the cache line corresponding
   to the address exists in the cache (using Set and Tag). If exists, then memory hit
   is predicted, otherwise, we predict a miss.
2. In case that the access is to symbolic memory location then the *AccessList* of the
   corresponding memory allocation-site is being checked and the accessed address
   is being compared to the ordered *AccessList* nodes as shown in the pseudocode
   at Fig. 5.9. At the first stage, the *Set* range of the addresses is being compared. If
   Set comparison was satisfied then the Tag is being checked. If, at the ordered
   search, a Set was satisfied but Tag was not, then we stop the search as cache line
   was evicted (other address with the same Set was accessed). In case that the Set
   was satisfied and then the Tag was satisfied then we assume a hit, otherwise, we
   assume that the access missed the cache.

5.4.3 Symbolic Execution Engine

The SEEM systems uses the KLEE [1] symbolic execution engine to execute the
programs symbolically. KLEE is a symbolic execution virtual machine that operates on the intermediate-representation of the LLVM (LLVM IR) assembly language as created by the LLVM compiler.

As shown in Fig. 5.7, the program (C, C ++ or other languages) is being
compiled with the LLVM compiler to generate LLVM bytecode (*.bc). The user
has the option to apply optimization on the bytecode with LLVM optimizer tool
(opt). The bytecode is the input for the KLEE system, which was extended with the
energy model and cache emulator.

Fig. 5.9 Cache hit/miss
prediction

```
IsCacheMIss_HIT(accessList,INDX){
  For(i=accessList.length; i>=0; i--){
  /* check if the Set field of accessed address matches
  (satisfiable) the Set field at accessList nodes */
      If(SAT(Access_i.index[a₃:a₂] ==  INDX[a₃:a₂])){
      /* check if Tags matches (satisfiable) */
          If(SAT(Access_i.index[a₅:a₄] ==  INDX[a₅:a₄])){
              return HIT;
          }else{
      /* if Set matches but Tag doesn't match then the
      cache line was evicted, stop searching ,return MISS */
              return MISS
          }
        }
      }
  return MISS
}
```

5.5 Results

The SEEM systems were tested on a 3rd Generation Intel® Core™ code name
IvyBridge that is described in Table 5.1.

Two sets of benchmarks were evaluated, the first set is syntactic benchmarks that
illustrates our method, validate it, and shows its accuracy. The second set of
benchmarks are real programs from the GNU COREUTILS [17] benchmarks.
Running on real benchmarks, we have demonstrated our tool with three compilation
flags: unrolling (unroll factor of 2), inlining, and unswitching, these runs shows
how maximums energy and power metrics change with different compilation flags
and enable the user to choose the combination that minimizes these metrics.

Table 5.1 System setup parameters

Processor	Intel® Core™ i7-3940XM (IvyBridge). 3 GHz fixed frequency. A quadcore with 8 MB L3 cache. Hardware prefetchers enabled
Memory	8 GB DDR3 @1600 MHz
OS	Ubuntu 14.04 64-bit
Benchmarks	GNU COREUTILS [17]
Compiler	LLVM compiler [15]

5.5.1 Simple Examples

We have evaluated three programs with the SEEM tool. Figure 5.10 shows the principal functions of these programs. The three programs execute memory copy from one array to the other, with different strides and iteration counts at each program. The arrays are global to the functions, and the allocation of the memory is done at the *main* function. The energy estimated by SEEM tool was compared to the Intel proprietary power simulator, which was used also for the energy model generation as explained in Sect. 5.4.1. At all runs, the hardware prefetching are disabled at the simulator.

The first program does memory copy from source array to destination array. The location from which the copy at the source is done depends on the stride value (argument to the function). We defined the stride parameter as symbolic value. The tool is able to predict the energy for the different cases of the symbolic stride. At the first experiment, we show the accuracy of our tool while applying loop unrolling on the program (a). Figure 5.11 shows the core energy results of SEEM compared to the power simulator data (Sim). The results are for concrete stride value (stride = 1) and various unrolling factor. We can see that applying unrolling reduces the energy, as less instructions are being executed (less compares, branches, and additions), the error rate compared to the simulator ranges between 5.1 and 10.4%. At this run, the cache power is mostly constant across the runs.

Fig. 5.10 Functions of the evaluated programs

```
int mem_copy_a(unsigned stride) {
    for(int i=0,j=0;i<N;j++,i+=stride){
        dst[j] = src[i];
    }
    return 0;
}
```
(a) Number of iteration is a function of stride

```
int mem_copy_b(unsigned stride) {
    int i,j;
    for(i=0,j=0;j<M;j++,i+=stride){
        dst[i] = src[i];
    return 0;
}
```
(b) Number of iterations is fixed (M)

```
int strides[8] = {32,8,16,4,128,1,64,2} ;
int mem_copy_c(unsigned stride_index) {
    int i,j;
    for(i=0,j=0;j<M;j++,i+=strides[stride_index]){
        dst[i] = src[i];
    }
    return 0;
}
```
(c) Number of iterations is fixed, stride is random

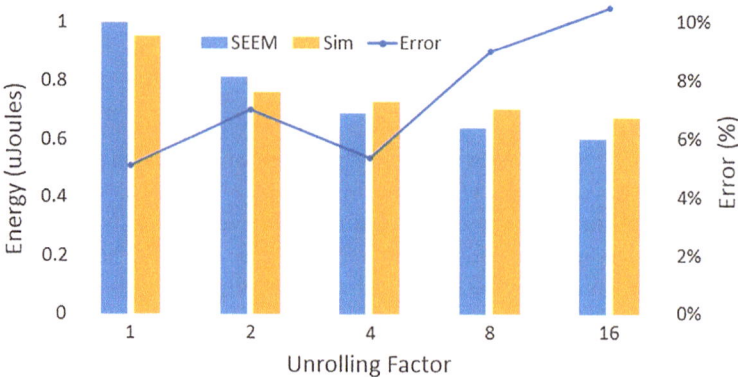

Fig. 5.11 Core energy result of running program (**a**) with concrete stride and various unrolling factors

At the second experiment, we ran program (b) with our tool. This program copies source array to destination array, the number of elements that are being copied is constant (M), and the array's indices are chosen by the stride parameter. The stride, in this case, affects the number of cache misses. For example, assuming that the cache line is 64 bytes, with a stride of 1 we will have a cache miss followed by 63 cache hits. For the case that the stride of 64 we will have cache miss every array access. At our configuration, the cache line is 64 bytes.

Figure 5.12 shows the energy consumption of the core and the cache for both the SEEM and power simulator. In addition, two lines are showing the error rate of SEEM compared to the simulator. We can see that the energy of the core and the cache is affected by the different stride values, despite that the number of the copied items is the same at all runs. The highest energy consumption occurs in the cases of 64 and 128 strides, as at these strides we have cache miss every memory access for both the source and destination arrays.

The core energy is affected by the cache miss mainly because of the cache penalty. As a delay is associated with the cache miss, at the time of miss the out of order code can run other instruction and utilize the Instruction-Level Parallelism (ILP), but at some stage no instructions are available for execution and the core is stalled. At stall time the core can clock gate some of the internal clocks at the front end, out of order, and execution clusters to save dynamic power, but the core will continue to consume leakage power (clusters are not power gated as the delay is relatively short) that is translated to additional energy at the stall time. The case is similar at the cache, part of the clocks are being gated to save dynamic power while waiting for the cache line to arrive from the upper cache at the memory hierarchy, but the cache will continue to consume leakage power at the stall cycles.

At the third program (c), we ran SEEM with symbolic value for the *stride_index* parameter, the *stride_index* is an index into array of strides (values of 1, 2, 4... 128 at random order), at this run SEEM was able to estimate the energy of all the paths, at this run, we have nearly the same result as shown in Fig. 5.12 (as the same strides

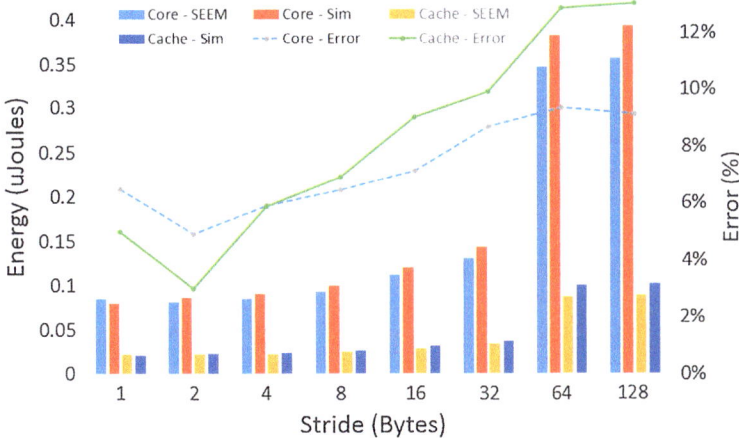

Fig. 5.12 Core energy results of running program (**b**) with various strides

were practically used). The purpose of this experiment is to show the effectiveness of the symbolic execution method in analyzing the energy of the program. This example cannot be handled with static analysis methods and tools (e.g. dataflow analysis or pattern detection). The alternative approach to our method is to write a "test plan" for checking the different paths and executing each case separately at real system. This process needs reasonable understanding of the programs and its flows, while with the SEEM tool all feasible paths are being estimated, and the user has the option to exclude non-interesting cases by adding constraints to the symbolic values.

5.5.1.1 Accuracy of SEEM

While the Intel power simulator has accuracy of 3–5% for performance workloads, Fig. 5.12 shows that the error rate at these runs is ranged between 3.1 and 12.9%. We can see that the error rate is higher when we have higher miss rate. There are few elements that causes this inaccuracy. One of these elements is the internal power features of the core, e.g., clock gating, these features are modeled inside the power simulator while at SEEM we did not model them as we are not modeling the microarchitecture behavior of the core or the cache. Another element is the out of order machine behavior, the SEEM runs the instruction in order, and assumes constant cycles per instructions (CPI), on the other hand the simulator runs the instructions out of order and can potentially run several instructions at the same time. Moreover, the simulator takes into account the over-all miss penalty (from L1 miss until memory), which is variable as it depends on the location of the data (L2, L3, and memory), but at SEEM case we assume fixed miss penalty. The instructions cache power is other aspect, at SEEM we have not modeled yet the instruction cache, hence variations on the power can exist, this part is more relevant to the first experiment with the unrolling. In addition, core microarchitecture features that exist at modern processors and not modeled at SEEM like the speculate-execution and prefetching affects the accuracy of our model for both the core and the cache power and energy estimations.

Dynamic power and energy estimation methods based on hardware energy reporting counters [21] or based on performance counters [28] have higher accuracy than our static method at SEEM, the accuracies of dynamic methods are about 5–7%. The higher accuracy is achieved mainly due the use of microarchitectural events (e.g. at [28]) or by the online reporting of the power consumption that is measured by physical hardware components (e.g. current sensors).

5.5.2 Bounding Heuristics

In Sect. 5.2 we explained that one of the problems at symbolic execution is the path explosion problem. Solutions to the path explosion problem generally use either heuristics for path-finding to increase code coverage [12], bounding the execution time of a single path, bounding the range of symbolic values or reduce execution time by parallelizing independent paths. In this section, we examine the effect of using execution time bounding and symbolic values bounding on the tools accuracy.

5.5.2.1 Execution Time Bounding

In this experiment, we ran program (b) several times with our tool, each run is with different limit on the execution time (max-time option at command-line), Fig. 5.13 summarizes the results of this run. It can be observed that with short bounds on the maximum execution time the maximum estimated energy by the tool is far from the actual maximum for both the core and the cache and the error is high (60–70%), while when enabling higher maximum execution time bound then, we see that the estimation is growing toward the actual maximum energy and the error is reducing (6–11%). This behavior is due to the fact that when enabling higher maximum execution time, then the symbolic execution engine is able to cover more values of the symbolic variables (Stride) and reach the value that maximizes the energy (Stride >= 64).

5.5.2.2 Bounding Symbolic Values

Similarly, to time bounding, in this experiment, we have bounded the symbolic value of program (b), running the program with different bounds resulted in

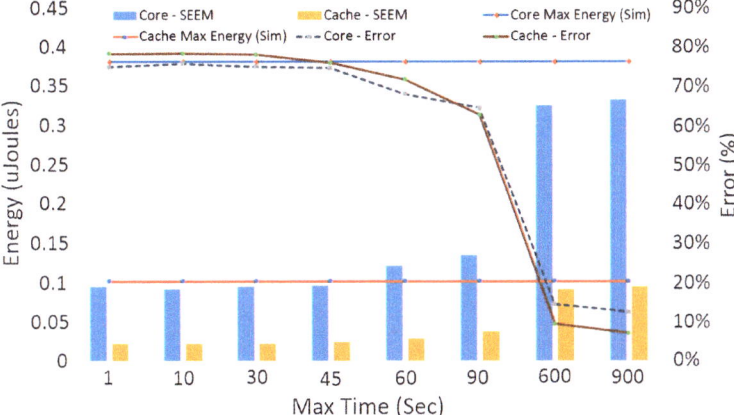

Fig. 5.13 SEEM max-energy estimation for program (**b**) with different time bounds (max-time)

different maximum energy estimation by our tool as shown in Fig. 5.14. Bounding the stride to 64, for example, will result in up to four times higher cache misses than the case with bounding it to 16.

5.5.3   Real Programs

We have evaluated our SEEM tool on real programs. We ran the tool on 89 programs of GNU COREUTILS [17] version 6.10, which contains roughly 80,000 lines of library code and 61,000 lines in the actual utilities. These programs interact extensively with their environment to provide a variety of functions, including managing the file system (e.g., ls, dd, and chmod), displaying and configuring system properties (e.g., logname, printenv, and hostname), controlling command invocation (e.g., nohup, nice, and env), processing text files (e.g., sort, od, and patch), and so on. They form the core user-level environment installed on many Unix systems.

5.5.3.1   Configuration and Heuristics

We ran the tool on all benchmarks while using symbolic values as parameters to each benchmark, we used the same setup as [1].

At this run, we used both time and symbolic values bounding heuristics. All benchmarks were ran using the following command

```
seem − −max-time 60 − −sym-args 10 2 2
− −sym-files 2 8 < benchmark-name>
```

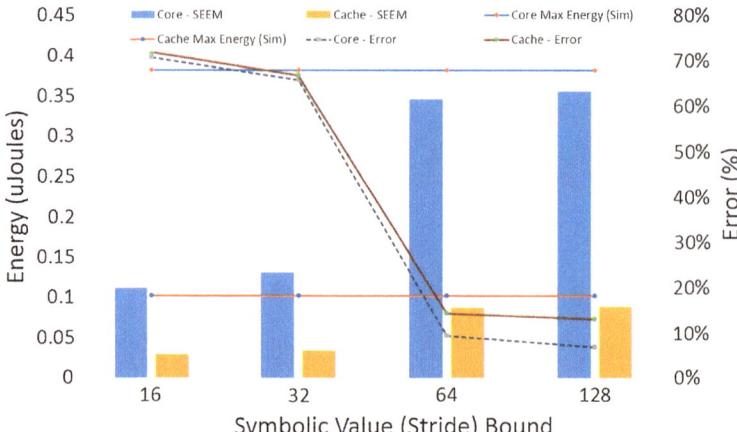

Fig. 5.14 SEEM max-energy estimation for program (**b**) with different bounds of symbolic value (stride)

As specified by the − −max-time option, we ran each tool for about 60 min (some finished before this limit). The other two options describe the symbolic inputs. The option − −sym-args 10 2 2 says to use zero to three command-line arguments, the first one character long, the others 10 characters long. The option − − sym-files 2 8 says to use standard input and one file, each holding eight bytes of symbolic data.

As explained in Sect. 5.4, the input program to SEEM tool is LLVM bytecode, the source code of each benchmark was first compiled to bytecode using the publicly available LLVM compiler [15]. The compilation done with three different flags that apply three different code optimizations (unrolling, inlining, and unswitching), the tool picked the optimization that minimizes the maximum power/ energy drawn at the core/cache by the programs.

5.5.3.2 Real Programs Results

Figure 5.15 shows the core and cache maximum energy behavior of part of the COREUTILS benchmarks as function of the code optimizations applied unrolling (factor of 2), inlining, or unswitching. In addition, Fig. 5.16 shows the core and cache maximum power behavior of part of the COREUTILS benchmarks as function of the code optimizations applied. These experiments show one of the usages of this tool which is the power and energy-aware complication. Figure 5.15 shows that the different compiler optimizations have different impact on energy and power of the core and cache, for example, the "dir" program showed improvement on energy consumption with all optimization at core and cache domain, while the "sort" program showed energy losses at the unswitching optimization for both the core and cache.

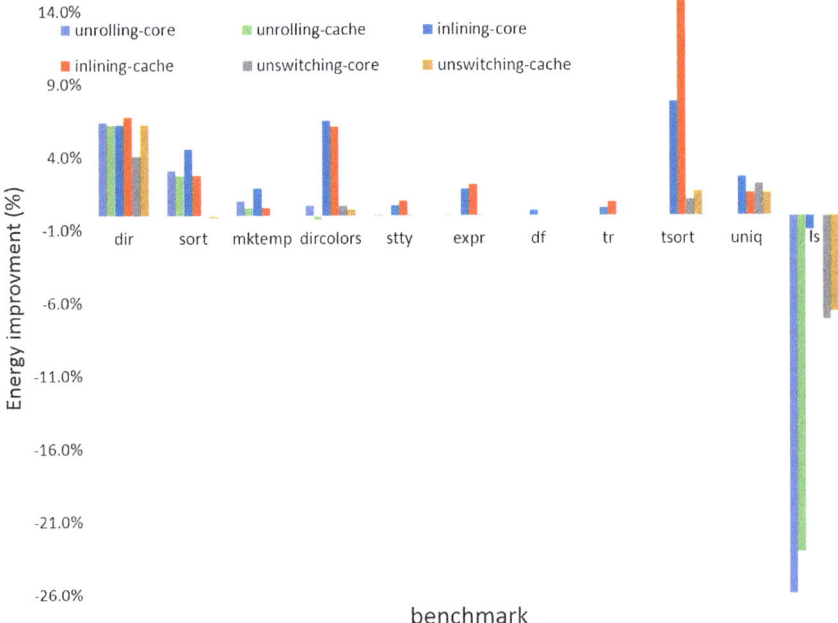

Fig. 5.15 Maximum energy behavior for the core and cache as function of compiler optimization (unrolling, inlining, and unswitching) relative to baseline

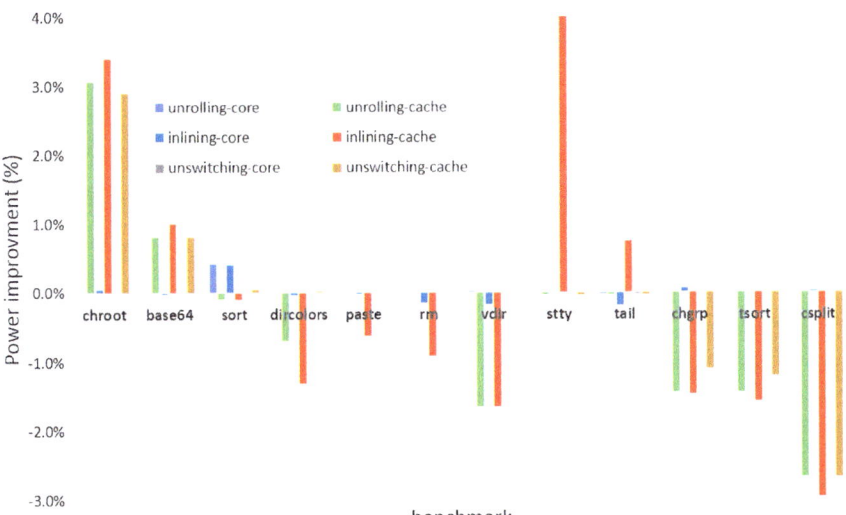

Fig. 5.16 Maximum power behavior for the core and cache as function of compiler optimization (unrolling, inlining, and unswitching) relative to baseline

The results for the rest of the benchmarks are not shown, as their gain/losses from the optimizations is nearly zero percent.

5.6   Discussion

In this work, we presented a tool that uses symbolic execution to estimate energy consumption of programs, the tool models the core and cache energy and doesnot take into account energy of other components at the System on Chip (SoC) and system level. In addition, the modeling was done for performance benchmarks and didnot take into account the various power state of the core (core C-states) and system (package C-states). In this section, we will explain how these aspects can be modeled and explain the challenges for supporting this.

5.6.1   Modeling Additional SoC Components

The IvyBridge processor includes several subsystems, in addition to the core and L1 cache, such as memory subsystems (ast-Level-Cache, memory-controller, and memory-IO), display subsystems (display engine, display-IO), graphics subsystem, in addition to several debug components. Moreover, at system level, components like Dynamic random-access memory (DRAM), Solid-State Drive (SSD), and network card are not modeled by our tool.

Part of the SoC and system components are not affected directly by the code that the user is running on the core; for example, the display power is mostly affected by the resolution setting, number of displays activated, and other display settings; on the other hand, components like memory subsystem, DDR, and SSD are affected directly by the running code on the cores and graphics engine.

5.6.1.1   Code Independent Energy

For modeling the power of components that are not affected by the user program, we can add a mapping between system configurations to constant power tax that these components will consume, multiplying the power by the time it takes to run the program will give a good estimation of the energy consumed by these parts of the system. Here, we need to take into account also the power states of these components, for example, the display has a power saving feature that reduces the power in case of static image of the screen.

5.6.1.2 Code Dependent Energy

Energy of system components that are affected by the code running on the core can be estimated by assigning energy per operation to each component that we want to model. In this work, we have implemented the L1 cache modeling, and we differentiated between two cases of the case—hit/miss. Modeling the L2, L3, Memory-Controller, DDR, and SSD is possible using the same method that we used at the L1 cache emulator.

5.6.2 Modeling Low Power States

Most of the SoC components have various power states, where each component can enter into these power states autonomously or by driver/operating system activation. For example, the cores have clock gating that is enabled autonomously by each subunit, in addition, cores have power state called C-states [20], these states are triggered by the operating system algorithm that chooses the level of power state to request based on core idleness period, in addition, the processor firmware is responsible to enter the system to system level idle-state called package-cstate based on the state on the core, graphics, and other SoC components.

In our tool, we donot have the system-level visibility in order to predict when the cores or system will enter idle power state. At this study, the assumption is that the applications that we are handling are performance applications that donot have idleness periods, this assumption means that most of the time the core is at running state (not idle), and this is aligned with the configuration where we measured the EPI for our power model.

5.6.3 Optimal Compiler Settings for Energy Reduction

Our tool is necessarily not finding the optimal energy or power, SEEM statically estimates the energy and power among all program feasible paths, this enables exploring the effect of different compilation setting on the power and energy of the program at compile-time. Finding the best compiler flag's settings for optimal energy or power is a hard problem that is not addressed in this work. Similar to optimization for performance, finding such a best set of optimization flags is a challenging task. This problem is NP-complete and thus, it is not possible to find the general exact solution. Moreover, prior art [29–31] shows that performance of different execution paths in a program is best optimized by different sequences of compiler flag's settings, this observation is correct also for energy and power optimization, we kept dealing with these issue in future work.

5.7   Related Works

At [19], the authors presented energy-aware programming support via symbolic execution. For each code path explored by a symbolic execution toolkit, the base energy cost of the core can be highlighted to the programmer. The methods require running the selected paths on the real machine and measuring the power. Our method is a static method that gives energy estimation on the core and cache, this method can be further extended to other CPU components. The method at [19] was evaluated on basic "test application" while our method was applied to a set of real benchmarks. Moreover, the processors that were checked at [19] is relatively old in-order execution machine while our work was done with the state of the art—out of order execution—Intel's processor.

Abstract interpretation-based cache analyses were used at [23] to calculate the worst-case-execution time (WCET), these methods capture an upper-bound of the cache state before every basic block in the application. This provides guarantees on the absence or presence of blocks in the cache. More precise approaches [22] rely on more costly models, e.g., to distinguish different contexts and timings for a same basic block. Our framework, however, uses symbolic execution and cache emulator to predict when a given memory access hits or misses the cache.

In [24], the authors attempt to perform static worst-case energy consumption (WCEC) analysis for a simple embedded processor, the Cortex M0+. This analysis is an absolute energy model, an energy model that provides the maximum energy consumption of each instruction. The authors argue that they can retrieve a safe bound. However, this is demonstrated on a single benchmark, bubble sort, only. The bound is 19% above a single hardware measurement; the authors acknowledge that this approach leads to over-approximations. Our framework, however, uses symbolic execution to estimate the energy on all paths and collect statistics on energy/power and not just WCEC.

Energy and power estimation has been studied for many years. Tiwari et al. were the first to propose the Instruction-Level Power Analysis (ILPA) approach for software power consumption modeling [2], which has been popular in recent years. They use a simple equation that divides the power cost of each instruction into the base cost and inter-instruction effects, including circuit state overhead, resource constraints and effect of cache misses. Each instruction and inter-instruction effect is annotated with a fixed value. Recent works showed energy estimation in higher level, Neville et al. [14] executed a static energy estimation at intermediate-representation of the LLVM [15] compiler (LLVM IR). The estimation didnot include the cache.

Other power estimation models based on cycle-level simulation. These tools use microarchitectural simulators to evaluate system performance and with the help of analytic power models to estimate consumption for each component of the platform. Wattch [26] and SimplePower [27] are examples of available tools. The power consumption of the main internal units is estimated using power macromodels, produced from lower level characterizations. The contributions of the unit activities are calculated and added together during the execution of the program on the cycle-accurate microarchitectural simulator. Though using cycle-accurate simulators has allowed accurate power estimation, evaluation and simulation time are very significant for the off-the-shelf processor. These tools can require thousands of instructions to simulate one processor's instruction, which is too slow to use for simulating modern interactive applications.

For dynamic power and energy estimation and optimization, various program characterization techniques, used normally for performance evaluation, should be used. Feature-aware techniques appear in many common practices in program characterization, e.g., for the classification and construction of workloads [30]. In addition, various intelligent machine learning-aided forms of profile-guided optimizations, e.g., selection of sequences of compiler optimizations and/or their settings [31] and runtime environment tuning. Moreover, analytic evaluation is used to evaluate performance of a computer system with respect to various system parameters and system workloads (e.g., various benchmarks and their inputs [29]). The auto-tuning technique can be applied for dynamic optimization, Collective mind [32] is a framework that provides an abstraction layer for auto-tuning. It tackles the chaos between different hardware, operating systems, software tools, search algorithms, and compilers versions, by allowing researchers to crowdsource information and cross-validate their findings.

In [33], an energy-aware compilation framework is presented that estimates and optimizes energy consumption of a given code, taking as input the energy/performance constraints, architectural and technological parameters, and energy models. The energy consumption in this work has been modeled for datapath, clock network, buses, caches, and main memory. Some of the application dependent parameters extracted from the code using the compiler, such as datapath accesses, number of execution cycles, bus transactions, cache misses, and memory transactions. The work focused on a single-issue, five-stage processor. This technique is much more complicated than our technique as it requires modeling many subcomponents of the processor, and it was illustrated on relatively simple processor, while our method is simpler and was illustrated on state of the art Intel processor.

5.8 Conclusions

In this work, we built the SEEM tool. A tool that uses symbolic execution to estimate energy and power consumption of programs. We demonstrated how the symbolic execution can be used to estimate energy of the core and cache at the various paths of the program. By running the programs with symbolic values, a prediction of the various run cases can be captured, which gives the software developer better visibility on his code. In addition, our tool enabled power and energy-aware compilation by running the program symbolically and picking the compilation flags that minimize the power or performance.

The SEEM tool can be enhanced further to give different statistics of the energy and power of the program, such as path with maximum energy, highest energy at given instruction window, highest power that can be drawn by any path, average power estimation, etc.

References

1. Cadar, Cristian, Daniel Dunbar, and Dawson R. Engler. "KLEE: Unassisted and Automatic Generation of High-Coverage Tests for Complex Systems Programs." OSDI. Vol. 8. 2008.
2. V. Tiwari, et al., "Instruction Level Power Analysis and Optimization of Software," in Journals of VLSI Signal Processing Systems, 1996, pp. 223–233.
3. King, James C. "Symbolic execution and program testing." Communications of the ACM 19.7 (1976): 385–394.
4. M. Popovich, A. V. Mezhiba, and E. G. Friedman, Power Distribution Networks With On-Chip Decoupling Capacitors. New York: Springer, 2008.
5. Reddi, Vijay Janapa, et al. "Voltage smoothing: Characterizing and mitigating voltage noise in production processors via software-guided thread scheduling." Microarchitecture (MICRO), 2010 43rd Annual IEEE/ACM International Symposium on. IEEE, 2010.
6. Reddi, Vijay Janapa, et al. "Voltage emergency prediction: Using signatures to reduce operating margins." High Performance Computer Architecture, 2009. HPCA 2009. IEEE 15th International Symposium on. IEEE, 2009.
7. McCurdy, Collin, Gabriel Marin, and Jeffrey S. Vetter. "Characterizing the impact of prefetching on scientific application performance." High Performance Computing Systems. Performance Modeling, Benchmarking and Simulation. Springer International Publishing, 2014. 115–135.
8. Lefurgy, Charles R., et al. "Active management of timing guardband to save energy in POWER7." proceedings of the 44th Annual IEEE/ACM International Symposium on Microarchitecture. ACM, 2011.
9. Haj-Yihia, Jawad, et al. "Compiler-directed power management for superscalars." ACM Transactions on Architecture and Code Optimization (TACO) 11.4 (2015): 48.
10. Een, Niklas, and Niklas Sörensson. "MiniSat: A SAT solver with conflict-clause minimization." Sat 5 (2005).
11. Anand, Saswat, Patrice Godefroid, and Nikolai Tillmann. "Demand-driven compositional symbolic execution." Tools and Algorithms for the Construction and Analysis of Systems. Springer Berlin Heidelberg, 2008. 367–381.
12. Ma, Kin-Keung, et al. "Directed symbolic execution." Static Analysis. Springer Berlin Heidelberg, 2011. 95–111.
13. Staats, Matt, and Corina Păsăreanu. "Parallel symbolic execution for structural test generation." Proceedings of the 19th international symposium on Software testing and analysis. ACM, 2010.
14. Grech, Neville, et al. "Static analysis of energy consumption for LLVM IR programs." Proceedings of the 18th International Workshop on Software and Compilers for Embedded Systems. ACM, 2015.
15. Lattner, Chris, and Vikram Adve. "LLVM: A compilation framework for lifelong program analysis & transformation." Code Generation and Optimization, 2004. CGO 2004. International Symposium on. IEEE, 2004.
16. Shao, Yakun Sophia, and David Brooks. "Energy characterization and instruction-level energy model of Intel's Xeon Phi processor." Proceedings of the International Symposium on Low Power Electronics and Design. IEEE Press, 2013.
17. Coreutils. www.gnu.org/software/coreutils.

18. Borghesi, Andrea, et al. "Power Capping in High Performance Computing Systems." Principles and Practice of Constraint Programming. Springer International Publishing, 2015.
19. Hönig, Timo, et al. "SEEP: exploiting symbolic execution for energy-aware programming." Proceedings of the 4th Workshop on Power-Aware Computing and Systems. ACM, 2011.
20. Intel 64 and IA-32 Architectures Software Developer's Manual, Volume 3, Section 14.9 (as of August 2014).
21. David, H., Gorbatov, E., Hanebutte, U.R., et al.: RAPL: memory power estimation and capping. In: 2010 ACM/IEEE International Symposium on Low-Power Electronics and Design (ISLPED), pp. 189–194. IEEE (2010).
22. S. Chattopadhyay and A. Roychoudhury. Scalable and precise refinement of cache timing analysis via path-sensitive verification. Real-Time Systems, 49(4), 2013.
23. C. Ferdinand and R. Wilhelm. Efficient and precise cache behavior prediction for real-time systems. Real-Time Systems, 17(2–3):131–181, 1999.
24. Wagemann, Peter, et al. "Worst-case energy consumption analysis for energy-constrained embedded systems." Proceedings of the 27th Euromicro Conference on Real-Time Systems (ECRTS). IEEE, 2015.
25. Hönig, Timo, et al. "Proactive energy-aware programming with PEEK." 2014 Conference on Timely Results in Operating Systems (TRIOS 14). USENIX, 2014.
26. Brooks, David, and Vivek Tiwari. "Wattch: A Framework for Architectural-Level Power Analysis and Optimizations." (2000).
27. W. Ye, N. Vijaykrishnam, M. Kandemir, and M. Irwin. The design and use of simplepower: a cycle accurate energy estimation tool. In Proc. Design Automation Conference DAC'00, June 2000.
28. Haj-Yihia, J., Yasin, A., Asher, Y.B. and Mendelson, A., 2016. Fine-grain power breakdown of modern out-of-order cores and its implications on Skylake-based systems. ACM Transactions on Architecture and Code Optimization (TACO), 13(4), p. 56.
29. Chen, Yang, Yuanjie Huang, Lieven Eeckhout, Grigori Fursin, Liang Peng, Olivier Temam, and Chengyong Wu. "Evaluating iterative optimization across 1000 datasets." ACM Sigplan Notices 45, no. 6 (2010): 448–459.
30. Eeckhout, Lieven, Hans Vandierendonck, and Koenraad De Bosschere. "Workload design: Selecting representative program-input pairs." In Parallel Architectures and Compilation Techniques, 2002. Proceedings. 2002 International Conference on, pp. 83–94. IEEE, 2002.
31. Cammarota, Rosario, Alexandru Nicolau, Alexander V. Veidenbaum, Arun Kejariwal, Debora Donato, and Mukund Madhugiri. "On the determination of inlining vectors for program optimization." In International Conference on Compiler Construction, pp. 164–183. Springer Berlin Heidelberg, 2013.
32. Fursin, Grigori, Renato Miceli, Anton Lokhmotov, Michael Gerndt, Marc Baboulin, Allen D. Malony, Zbigniew Chamski, Diego Novillo, and Davide Del Vento. "Collective mind: Towards practical and collaborative auto-tuning." Scientific Programming 22, no. 4 (2014): 309–329.
33. Kadayif, Ismail, M. Kandemir, Guilin Chen, Narayanan Vijaykrishnan, Mary Jane Irwin, and Anand Sivasubramaniam. "Compiler-directed high-level energy estimation and optimization." ACM Transactions on Embedded Computing Systems (TECS) 4, no. 4 (2005): 819–850.
34. Gustaffson, Jan, Andreas Ermedahl, and Björn Lisper. "Algorithms for infeasible path calculation." In OASIcs-OpenAccess Series in Informatics, vol. 4. Schloss Dagstuhl-Leibniz-Zentrum für Informatik, 2006.
35. Lam, Patrick, Eric Bodden, Ondrej Lhoták, and Laurie Hendren. "The Soot framework for Java program analysis: a retrospective." In Cetus Users and Compiler Infastructure Workshop (CETUS 2011), vol. 15, p. 35. 2011.

Printed by Printforce, the Netherlands